Lecture Notes in Mathematics 2197

More information about this series at http://www.springer.com/series/304

Saint-Flour Probability Summer School

The Saint-Flour volumes are reflections of the courses given at the Saint-Flour Probability Summer School. Founded in 1971, this school is organised every year by the Laboratoire de Mathématiques (CNRS and Université Blaise Pascal, Clermont-Ferrand, France). It is intended for PhD students, teachers and researchers who are interested in probability theory, statistics, and in their applications.

The duration of each school is 13 days (it was 17 days up to 2005), and up to 70 participants can attend it. The aim is to provide, in three high-level courses, a comprehensive study of some fields in probability theory or Statistics. The lecturers are chosen by an international scientific board. The participants themselves also have the opportunity to give short lectures about their research work.

Participants are lodged and work in the same building, a former seminary built in the 18th century in the city of Saint-Flour, at an altitude of 900 m. The pleasant surroundings facilitate scientific discussion and exchange.

The Saint-Flour Probability Summer School is supported by:
- Université Blaise Pascal
- Centre National de la Recherche Scientifique (C.N.R.S.)
- Ministère délégué à l'Enseignement supérieur et à la Recherche

For more information, see

http://recherche.math.univ-bpclermont.fr/stflour/stflour-en.php

Christophe Bahadoran
bahadora@math.univ-bpclermont.fr

Arnaud Guillin
Arnaud.Guillin@math.univ-bpclermont.fr

Laurent Serlet
Laurent.Serlet@math.univ-bpclermont.fr

Université Blaise Pascal – Aubière cedex, France

Sourav Chatterjee

Large Deviations
for Random Graphs

École d'Été de Probabilités de Saint-Flour
XLV - 2015

 Springer

Sourav Chatterjee
Department of Statistics
Stanford University
Stanford, CA, USA

ISSN 0075-8434   ISSN 1617-9692   (electronic)
Lecture Notes in Mathematics
ISSN 0721-5363
École d'Été de Probabilités de Saint-Flour
ISBN 978-3-319-65815-5   ISBN 978-3-319-65816-2   (eBook)
DOI 10.1007/978-3-319-65816-2

Library of Congress Control Number: 2017951112

Mathematics Subject Classification (2010): 60-XX

Printed on acid-free paper

This Springer imprint is published by Springer Nature
The registered company is Springer International Publishing AG
The registered company address is: Gewerbestrasse 11, 6330 Cham, Switzerland

To the memory of my grandfather, Tarapada Chatterjee

Preface

These lecture notes were prepared for the 45th Saint-Flour Probability Summer School in July 2015. They contain an exposition of the main developments in the large deviations theory for random graphs in the last few years. I have tried to make the exposition as self-contained as possible, so that the reader will have essentially no need for looking up external results. For example, the necessary components of graph limit theory are developed from scratch. Instead of going through Szemerédi's regularity lemma, I have taken an alternative analytic approach that develops the discrete and continuous theory "at one go", making it unnecessary to use martingales for passing from discrete to continuous. Similarly, the required results from classical large deviations theory and concentration of measure are also developed from scratch.

After the above preparatory materials, the main topics covered here are large deviations theory for dense random graphs, exponential random graph models, nonlinear large deviations, and large deviations for sparse random graphs.

I have tried to write the monograph in a way that is accessible to beginning graduate students in mathematics and statistics, with some background in graduate-level probability theory. Advanced readers may find some parts of the exposition to be too elementary.

To avoid clutter, references to the literature are not given within the main material but summarized at the end of each chapter. I have tried to be as comprehensive as possible in my literature review, and I apologize for any inadvertent omission.

It was a matter of great honor for me to be invited to deliver these lectures, which have a legendary status in the world of probability theory. I thank the scientific committee for selecting me for this honor—and, moreover, for allowing me to delay my lectures by 1 year after I had to cancel in 2014 due to the birth of my son. I am especially grateful to the local organizers, Laurent Serlet and Christophe Bahadoran, for taking care of every little practical detail during my visit.

It was a great pleasure and a great learning experience to interact with my co-lecturers, Sara van de Geer and Lorenzo Zambotti.

The students and other attendees at Saint-Flour were a source of inspiration. It is every lecturer's dream to have an audience of that caliber.

The development of large deviations for random graphs involved a number of my colleagues, who have contributed greatly to the topic at various points. I would like to take this opportunity to acknowledge their contributions. In particular, I would like to thank Bhaswar Bhattacharya, Amir Dembo, Shirshendu Ganguly, Eyal Lubetzky, Charles Radin, Raghu Varadhan, Mei Yin, and Yufei Zhao for the many useful interactions I have had with them over the last few years.

I am grateful to Alexander Dunlap for carefully reading the manuscript and pointing out many typos and mistakes. The responsibility for any error that still remains is solely mine.

Finally, a very special thanks to my wife and son for bringing joy and fulfillment to my life, to my parents for their good wishes, and to Persi Diaconis for his advice and encouragement.

Stanford, CA, USA   Sourav Chatterjee
June 2017

Contents

Chapter 1
Introduction

This introductory chapter lays out the general plan of the book and gives a quick description of the main issues, reproducing from the first three sections of my recent survey article [6].

1.1 Large Deviations

The theory of large deviations studies two things: (a) the probabilities of rare events, and (b) the conditional probabilities of events given that some rare event has occurred. Often, the second question is more interesting than the first, but it is usually essential to answer the first question to be able to understand how to approach the second.

As an illustration, consider the following simple example. Toss a fair coin n times, where n is a large number. Under normal circumstances, you expect to get approximately $n/2$ heads. Also, you expect to get roughly $n/4$ pairs of consecutive heads. However, suppose that the following rare event occurs: the tosses yield $\geq 2n/3$ heads. Classical large deviations theory allows us to compute that the probability of this rare event is

$$e^{-n\log(2^{5/3}/3)(1+o(1))} \tag{1.1.1}$$

as $n \to \infty$. Moreover, it can be shown that if this rare event has occurred, then it is highly likely that there are approximately $4n/9$ pairs of consecutive heads instead of the usual $n/4$.

So, how is the estimate (1.1.1) obtained? The argument goes roughly as follows. Let $X_1, \ldots, X_n$ be independent random variables, such that $\mathbb{P}(X_i = 0) = \mathbb{P}(X_i = 1)$

© Springer International Publishing AG 2017
S. Chatterjee, *Large Deviations for Random Graphs*, Lecture Notes in Mathematics 2197, DOI 10.1007/978-3-319-65816-2_1

$= 1/2$ for each i. Then the number of heads in n tosses of a fair coin can be modeled by the sum $S_n := X_1 + \cdots + X_n$. For any $\theta \geq 0$,

$$
\begin{aligned}
\mathbb{P}(S_n \geq 2n/3) &= \mathbb{P}(e^{\theta S_n} \geq e^{2\theta n/3}) \\
&\leq \frac{\mathbb{E}(e^{\theta S_n})}{e^{2\theta n/3}} \quad \text{(Markov's inequality)} \\
&= \frac{\mathbb{E}(\prod_{i=1}^n e^{\theta X_i})}{e^{2\theta n/3}} = \frac{\prod_{i=1}^n \mathbb{E}(e^{\theta X_i})}{e^{2\theta n/3}} \quad \text{(independence)} \\
&= e^{-2\theta n/3} \left(\frac{1 + e^\theta}{2} \right)^n .
\end{aligned}
$$

Optimizing over θ gives the desired upper bound. To prove the lower bound, take some $\epsilon > 0$ and define a random variable Z as:

$$
Z := \begin{cases} 1 & \text{if } 2n/3 \leq S_n \leq (1 + \epsilon)2n/3, \\ 0 & \text{otherwise.} \end{cases}
$$

Then for any $\theta \geq 0$,

$$
\begin{aligned}
\mathbb{P}(S_n \geq 2n/3) &\geq \mathbb{E}(Z) \\
&\geq e^{-\theta(1+\epsilon)2n/3} \mathbb{E}(e^{\theta S_n} Z) \quad \text{(since } S_n \leq (1 + \epsilon)2n/3 \text{ when } Z \neq 0) \\
&= e^{-\theta(1+\epsilon)2n/3} \left(\frac{1 + e^\theta}{2} \right)^n \frac{\mathbb{E}(e^{\theta S_n} Z)}{\mathbb{E}(e^{\theta S_n})} .
\end{aligned}
$$

The proof is completed by showing that if θ is chosen to be the same number that optimized the upper bound and ϵ is sent to zero sufficiently slowly as $n \to \infty$, then

$$
\frac{\mathbb{E}(e^{\theta S_n} Z)}{\mathbb{E}(e^{\theta S_n})} = e^{o(n)} .
$$

Establishing the above claim is the most nontrivial part of the whole argument, but is by now standard. This is sometimes called the 'change of measure trick'.

The above example has a built-in linearity, which allowed us to explicitly compute $\mathbb{E}(e^{\theta S_n})$. Generalizing this idea, classical large deviations theory possesses a collection of powerful tools to deal with linear functionals of independent random variables, random vectors, random functions, random probability measures and other abstract random objects. The classic text of Dembo and Zeitouni [10] contains an in-depth introduction to this broad area.

1.2 The Problem with Nonlinearity

In spite of the remarkable progress with linear functionals, there are no general tools for large deviations of nonlinear functionals. Nonlinearity arises naturally in many contexts. For instance, the analysis of real-world networks has been one of the most popular scientific endeavors in the last two decades, and rare events on networks are often nonlinear in nature. This is demonstrated by the following simple example.

Construct a random graph on n vertices by putting an undirected edge between any two with probability p, independently of each other. This is known as the Erdős–Rényi $G(n, p)$ model, originally defined by Erdős and Rényi [12]. The model is too simplistic to be a model for any real-world network, but has many nice mathematical properties and has led to the developments of many new techniques in combinatorics and probability theory over the years. One can ask the following large deviation questions about this model:

(a) What is the probability that the number of triangles in a $G(n, p)$ random graph is at least $1 + \delta$ times the expected value of the number of triangles, where δ is some given number?

(b) What is the most likely structure of the graph, if we know that the above rare event has occurred?

This is an example of a nonlinear problem, because the number of triangles in $G(n, p)$ is a degree three polynomial of independent random variables. To see this, let $\{1, \ldots, n\}$ be the set of vertices, and let X_{ij} be the random variable that is 1 if the edge $\{i, j\}$ is present in the graph and 0 if not. Then $(X_{ij})_{1 \le i < j \le n}$ are independent random variables, and the number of triangles is nothing but

$$\frac{1}{6} \sum_{i,j,k=1}^{n} X_{ij} X_{jk} X_{ki},$$

which is a polynomial of degree three. Until even a few years ago, large deviations theory did not have the tools to answer such basic questions about nonlinear functions of independent random variables, although a number of powerful concentration inequalities were available for computing upper and lower bounds on tail probabilities [15, 16, 21, 22].

1.3 Recent Developments

The large deviation theory for the Erdős–Rényi random graph was developed a few years ago in Chatterjee and Varadhan [9], taking to completion a program initiated in an unpublished manuscript of Bolthausen et al. (Large deviations for random matrices and random graphs, Private communication, 2003). The theory brought together ideas from classical large deviations theory and tools from combinatorics and graph theory, such as Szemerédi's regularity lemma and the theory of graph limits. The calculations dictated by the theory led to surprising conclusions, even in the simplest of applications such as the following. Let $T_{n,p}$ be the number of triangles in the Erdős–Rényi graph $G(n, p)$. What is the most likely structure of the graph if the rare event

$$E := \{T_{n,p} \geq (1 + \delta)\mathbb{E}(T_{n,p})\}$$

happens, where δ is a given positive constant? For instance, are all the extra triangles likely to be arising from a small subset of vertices with high connectivity amongst themselves? Or do they occur because the graph has an excess number of edges spread uniformly?

Surprisingly, the large deviation theory of [9] implies that both scenarios can happen. If p is smaller than a threshold, then there exist $0 < \delta_1 < \delta_2$ such that if $0 < \delta \leq \delta_1$ or $\delta \geq \delta_2$, then conditional on the event E, the graph behaves like $G(n, r)$ for some $r > p$; and if $\delta_1 < \delta < \delta_2$, then the conditional structure is *not* like an Erdős–Rényi graph. Explicit formulas for δ_1 and δ_2 were derived by Lubetzky and Zhao [17].

In other words, if the number triangles exceeds the expected value by a little bit or by a lot, then the most likely scenario is that there is an excess number of edges spread uniformly; and if the surplus amount belongs to a middle range, then the structure of the graph is likely to be inhomogeneous. There is probably no way that the above result could have been guessed from intuition; it was derived purely from a set of mathematical formulas.

The general theory of [9] and its main results and applications are described in Chaps. 5 and 6, after reviewing some preparatory materials in Chaps. 2, an introduction to graph limit theory in Chap. 3, and an introduction to classical large deviations theory in Chap. 4.

The large deviation theory for the Erdős–Rényi model has been extended to more realistic models of random graphs. For example, it was applied to exponential random graph models in Chatterjee and Diaconis [8] and a number of subsequent papers by Charles Radin, Mei Yin, Rick Kenyon and their collaborators [1, 13, 14, 19, 20, 23]. These models are widely used in the analysis of real social networks. Applications of random graph large deviations to exponential random graph models are discussed in Chap. 7.

The theory developed in [9] has one serious limitation: it applies only to dense graphs. A graph is called dense if the average vertex degree is comparable to the total number of vertices (recall that the number of neighbors of a vertex is called its degree). For example, in the Erdős–Rényi model with $n = 10000$ and $p = .3$, the average degree is roughly 3000. This is not true for real networks, which are usually sparse. Unfortunately, the graph theoretic tools used for the analysis of large deviations for random graphs are useful only in the dense setting. In spite of considerable progress in developing a theory of sparse graph limits [3–5], there is still no result that fully captures the power of Szemerédi's lemma in the sparse setting. In the absence of such tools, a nascent theory of 'nonlinear large deviations', developed in Chatterjee and Dembo [7] and recently improved by Eldan [11], has been helpful in solving some questions about large deviations for sparse random graphs, for example in Lubetzky and Zhao [18] and Bhattacharya, Ganguly, Lubetzky and Zhao [2]. This theory is discussed in Chap. 8.

References

1. Aristoff, D., & Radin, C. (2013). Emergent structures in large networks. *Journal of Applied Probability, 50*(3), 883–888.
2. Bhattacharya, B. B., Ganguly, S., Lubetzky, E., & Zhao, Y. (2015). Upper tails and independence polynomials in random graphs. *arXiv preprint arXiv:1507.04074.*
3. Bollobás, B., & Riordan, O. (2009). Metrics for sparse graphs. In *Surveys in combinatorics 2009*, vol. 365, London Mathematical Society Lecture Note Series (pp. 211–287). Cambridge: Cambridge University Press.
4. Borgs, C., Chayes, J. T., Cohn, H., & Zhao, Y. (2014). An L^p theory of sparse graph convergence I: Limits, sparse random graph models, and power law distributions. *arXiv preprint arXiv:1401.2906.*
5. Borgs, C., Chayes, J. T., Cohn, H., & Zhao, Y. (2014). An L^p theory of sparse graph convergence II: LD convergence, quotients, and right convergence. *arXiv preprint arXiv:1408.0744.*
6. Chatterjee, S. (2016). An introduction to large deviations for random graphs. *Bulletin of the American Mathematical Society, 53*(4), 617–642.
7. Chatterjee, S., & Dembo, A. (2016). Nonlinear large deviations. *Advances in Mathematics, 299*, 396–450.
8. Chatterjee, S., & Diaconis, P. (2013). Estimating and understanding exponential random graph models. *Annals of Statistics, 41*(5), 2428–2461.
9. Chatterjee, S., & Varadhan, S. R. S. (2011). The large deviation principle for the Erdős-Rényi random graph. *European Journal of Combinatorics, 32*(7), 1000–1017.
10. Dembo, A., & Zeitouni, O. (2010). *Large deviations techniques and applications.* Corrected reprint of the second (1998) edition. Berlin: Springer.
11. Eldan, R. (2016). Gaussian-width gradient complexity, reverse log-Sobolev inequalities and nonlinear large deviations. *arXiv preprint arXiv:1612.04346.*
12. Erdős, P., & Rényi, A. (1960). On the evolution of random graphs. *Publication of the Mathematical Institute of the Hungarian Academy of Sciences, 5*, 17–61.
13. Kenyon, R., Radin, C., Ren, K., & Sadun, L. (2014). Multipodal structure and phase transitions in large constrained graphs. *arXiv preprint arXiv:1405.0599.*
14. Kenyon, R., & Yin, M. (2014). On the asymptotics of constrained exponential random graphs. *arXiv preprint arXiv:1406.3662.*
15. Kim, J. H., & Vu, V. H. (2000). Concentration of multivariate polynomials and its applications. *Combinatorica, 20*(3), 417–434.
16. Latała, R. (1997). Estimation of moments of sums of independent real random variables. *The Annals of Probability, 25*(3), 1502–1513.
17. Lubetzky, E., & Zhao, Y. (2015). On replica symmetry of large deviations in random graphs. *Random Structures Algorithms, 47*(1), 109–146.
18. Lubetzky E., & Zhao, Y. (2017). On the variational problem for upper tails of triangle counts in sparse random graphs. *Random Structures Algorithms, 50*(3), 420–436.
19. Radin, C., & Sadun, L. (2013). Phase transitions in a complex network. *Journal of Physics A, 46*, 305002.
20. Radin, C., & Yin, M. (2011). Phase transitions in exponential random graphs. *arXiv preprint arXiv:1108.0649.*

21. Talagrand, M. (1995). Concentration of measure and isoperimetric inequalities in product spaces. *Publications Mathématiques de l'Institut des Hautes Études Scientifiques, 81,* 73–205.
22. Vu, V. H. (2002). Concentration of non-Lipschitz functions and applications. Probabilistic methods in combinatorial optimization. *Random Structures Algorithms, 20*(3), 262–316.
23. Yin, M. (2013). Critical phenomena in exponential random graphs. *Journal of Statistical Physics, 153*(6), 1008–1021.

Chapter 2
Preparation

Let $[0, 1]^d$ be the d-dimensional unit hypercube. The cases $d = 1$ and $d = 2$ are the relevant ones in this manuscript. This chapter summarizes some basic facts about $L^2([0, 1]^d)$. I will assume that the reader is familiar with Lebesgue measure, Borel sigma-algebra, integration, conditional expectation, basic results about integrals such as Fatou's lemma, monotone convergence theorem, dominated convergence theorem, Hölder's inequality and Minkowski's inequality, and elementary notions from topology such as the abstract definition of a topological space and the properties of continuous functions.

2.1 Probabilistic Preliminaries

This section summarizes some basic results from probability that are widely used in this monograph. Let $(\Omega, \mathscr{F}, \mathbb{P})$ be a probability space and $f : \Omega \to \mathbb{R}$ be a measurable function. Let R be an interval containing the range of f, and let $\phi : R \to \mathbb{R}$ be a convex function, meaning that for all $x, y \in R$ and $t \in [0, 1]$,

$$\phi(tx + (1 - t)y) \leq t\phi(x) + (1 - t)\phi(y).$$

Proposition 2.1 (Jensen's Inequality) *Let Ω, $\mathbb{P}$, ϕ and f be as above. Let*

$$m := \int_\Omega f(x) \, d\mathbb{P}(x).$$

Then

$$\phi(m) \leq \int_\Omega \phi(f(x)) \, d\mathbb{P}(x).$$

© Springer International Publishing AG 2017
S. Chatterjee, *Large Deviations for Random Graphs*, Lecture Notes
in Mathematics 2197, DOI 10.1007/978-3-319-65816-2_2

Moreover, if ϕ is nonlinear in every open neighborhood of m, then equality holds in the above inequality if and only if $f = m$ $\mathbb{P}$-almost surely.

Proof Note that $m \in R$. Since ϕ is a convex function, there exist $a, b \in \mathbb{R}$ such that $ax + b \leq \phi(x)$ for all $x \in \mathbb{R}$ and $am + b = \phi(m)$. To see this, simply observe that by the convexity of ϕ,

$$\liminf_{x \downarrow m} \frac{\phi(x) - \phi(m)}{x - m} \geq \limsup_{x \uparrow m} \frac{\phi(m) - \phi(x)}{m - x},$$

choose a to be a number between these two limits, and choose b to satisfy $ax + b = \phi(m)$. With these choices, the required properties follow from convexity. Having obtained a and b, observe that

$$\int_{\Omega} \phi(f(x)) \, d\mathbb{P}(x) \geq \int_{\Omega} (af(x) + b) \, d\mathbb{P}(x) = am + b = \phi(m),$$

which is the desired inequality. If ϕ is nonlinear in every open neighborhood of m, then the convexity of ϕ implies that with a suitable choice of a, it can be guaranteed that $\phi(x) > ax + b$ for all $x \neq m$. Thus, equality can hold if and only if $f = m$ $\mathbb{P}$-almost surely. $\square$

Let $(\Omega, \mathscr{F}, \mathbb{P})$ be a probability space. Recall that $L^2(\Omega, \mathscr{F}, \mathbb{P})$ is the space of all $\mathscr{F}$-measurable $f : \Omega \to \mathbb{R}$ such that

$$\|f\| := \left(\int_{\Omega} f(x)^2 \, dx \right) < \infty.$$

This is actually a normed space of equivalence classes, where two functions f and g are said to be equivalent if $\|f - g\| = 0$, which is the same as saying $f = g$ except possibly on a set of measure zero. We will, however, always treat the elements of this space as functions rather than as equivalence classes.

The most basic fact about L^2 is that it is a complete metric space. To prove this, we need two lemmas.

Lemma 2.1 (Chebychev's Inequality) *For any $f \in L^2(\Omega, \mathscr{F}, \mathbb{P})$ and $\epsilon > 0$,*

$$\mathbb{P}(\{x \in \Omega : |f(x)| \geq \epsilon\}) \leq \frac{\|f\|^2}{\epsilon^2}.$$

Proof Let g be a function which is 1 if $|f(x)| \geq \epsilon$ and 0 otherwise. Then $g \leq |f|/\epsilon$ everywhere. Therefore $\|g\|^2 \leq \|f\|^2/\epsilon^2$, which is exactly the statement of the lemma. $\square$

Lemma 2.2 (Borel–Cantelli Lemma) *If $\{A_n\}_{n\geq 1}$ is a sequence sets in $\mathscr{F}$ such that*

$$\sum_{n=1}^{\infty} \mathbb{P}(A_n) < \infty,$$

then

$$\mathbb{P}(\{x : x \in \text{infinitely many } A_n\text{'s}\}) = 0.$$

Proof Let $N(x) = $ the number of n such that $x \in A_n$. By the monotone convergence theorem,

$$\|\sqrt{N}\|^2 = \int_{[0,1]^d} N(x)\, dx = \sum_{n=1}^{\infty} \mathbb{P}(A_n) < \infty.$$

Therefore by Chebychev's inequality, $\mathbb{P}(\{x : N(x) \geq L\}) \to 0$ as $L \to \infty$, which proves that for $\mathbb{P}$-almost all $x \in \Omega, N(x) < \infty$. □

Proposition 2.2 (Completeness of L^2) *The space $L^2(\Omega, \mathscr{F}, \mathbb{P})$ is complete; that is, every Cauchy sequence converges to a limit. Moreover, any Cauchy sequence has a subsequence that converges $\mathbb{P}$-almost everywhere to its limit.*

Proof Let $\{f_n\}_{n\geq 1}$ be a Cauchy sequence. Then there exists a sequence $n_k \to \infty$ such that $\|f_{n_k} - f_{n_{k+1}}\| \leq 2^{-k}$ for each k. Let

$$A_k := \{x : |f_{n_k}(x) - f_{n_{k+1}}(x)| > 2^{-k/2}\}.$$

Then by Chebychev's inequality,

$$\mathbb{P}(A_k) \leq 2^{-k}.$$

Therefore by the Borel–Cantelli lemma, the set of all x such that $x \in$ infinitely many A_k's has measure zero. Now if $x \in$ only finitely many A_k's, then the sequence $\{f_{n_k}(x)\}_{k\geq 1}$ is a Cauchy sequence in $\mathbb{R}$. Let $f(x)$ denote the limit of this sequence. Then for each k, by Fatou's lemma and Minkowski's inequality,

$$\|f_{n_k} - f\| \leq \liminf_{l \to \infty} \|f_{n_k} - f_{n_l}\| \leq 2^{-k+1}.$$

The convergence of f_n to f now follows easily by the Cauchy property of the sequence $\{f_n\}_{n\geq 1}$. □

In the rest of this chapter, we will mostly specialize to $\Omega = [0,1]^d$, $\mathscr{F} = $ the Borel sigma-algebra generated by open sets, and $\mathbb{P} = $ Lebesgue measure.

2.2   Discrete Approximations of L^2 Functions

For each $k \geq 0$, let $\mathscr{D}_k$ be the set of all closed dyadic cubes of the form

$$\left[\frac{i_1 - 1}{2^k}, \frac{i_1}{2^k}\right] \times \left[\frac{i_2 - 1}{2^k}, \frac{i_2}{2^k}\right] \times \cdots \times \left[\frac{i_d - 1}{2^k}, \frac{i_d}{2^k}\right]$$

for some integers $1 \leq i_1, \ldots, i_d \leq 2^k$. Let

$$\mathscr{D} := \bigcup_{k=0}^{\infty} \mathscr{D}_k.$$

Suppose that a function $f \in L^2([0,1]^d)$ has the property that its integral over D is zero for every $D \in \mathscr{D}_k$. We will now show that such a function must be equal to zero almost everywhere. To prove this, we need a basic result from measure theory, known as the monotone class theorem.

Proposition 2.3 (Monotone Class Theorem) *Let Ω be any set and let $\mathscr{F}$ be an algebra of subsets of Ω. That is, $\Omega \in \mathscr{F}$, whenever $A \in \mathscr{F}$, $A^c := \Omega \setminus A$ is also in $\mathscr{F}$, and whenever $A, B \in \mathscr{F}$, $A \cup B$ is also in $\mathscr{F}$. Let $\mathscr{G}$ be the smallest collection of subsets of Ω such that $\mathscr{G} \supseteq \mathscr{F}$ and $\mathscr{G}$ is closed under monotone unions and intersection. That is, if $A_1 \subseteq A_2 \subseteq \cdots$ are members of $\mathscr{G}$, then $A_1 \cup A_2 \cup \cdots$ is also in $\mathscr{G}$, and if $B_1 \supseteq B_2 \supseteq \cdots$ are members of $\mathscr{G}$, then $B_1 \cap B_2 \cap \cdots$ is also in $\mathscr{G}$. Then $\mathscr{G}$ is a sigma-algebra.*

Proof Take any $A \in \mathscr{F}$, and let $\mathscr{G}_A$ be the set of all $B \in \mathscr{G}$ such that $A \cap B$, $A \cap B^c$ and $A^c \cap B$ are all in $\mathscr{G}$. Since $\mathscr{G}$ is closed under monotone unions and intersections, so is $\mathscr{G}_A$. Since $\mathscr{F}$ is an algebra and is contained in $\mathscr{G}$, therefore $\mathscr{F} \subseteq \mathscr{G}_A$. By the minimality of $\mathscr{G}$, this implies that $\mathscr{G}_A = \mathscr{G}$. Next, take any $B \in \mathscr{G}$ and any $A \in \mathscr{F}$. Since $\mathscr{G}_A = \mathscr{G}$, therefore $A \in \mathscr{G}_B$. Thus, $\mathscr{F} \subseteq \mathscr{G}_B$. Also, $\mathscr{G}_B$ is a monotone class. Therefore $\mathscr{G}_B = \mathscr{G}$. Since this is true for all $B \in \mathscr{G}$, this shows that $\mathscr{G}$ is an algebra. Since $\mathscr{G}$ is closed under monotone unions and intersections, $\mathscr{G}$ must be a sigma-algebra. □

Proposition 2.4 *Suppose that $f \in L^2([0,1]^d)$ satisfies*

$$\int_D f(x)\, dx = 0$$

for every $D \in \mathscr{D}$, where $\mathscr{D}$ is the collection of all closed dyadic cubes defined above. Then $f = 0$ almost everywhere.

Proof For each $k \geq 0$, let $\mathscr{D}'_k$ be the set of all half-open dyadic cubes of the form

$$\left[\frac{i_1 - 1}{2^k}, \frac{i_1}{2^k}\right) \times \left[\frac{i_2 - 1}{2^k}, \frac{i_2}{2^k}\right) \times \cdots \times \left[\frac{i_d - 1}{2^k}, \frac{i_d}{2^k}\right)$$

for some integers $1 \le i_1, \ldots, i_d \le 2^k$. Let

$$\mathscr{D}' := \bigcup_{k=0}^{\infty} \mathscr{D}'_k.$$

Let $\mathscr{F}$ be the set of all finite unions of elements of $\mathscr{D}'$. It is easy to see that $\mathscr{F}$ is an algebra of subsets of $[0, 1)^d$.

Let $\mathscr{G}$ be the set of all Borel subsets $B \subseteq [0, 1)^d$ such that

$$\int_B f(x)\, dx = 0.$$

By the dominated convergence theorem, $\mathscr{G}$ is closed under monotone unions and intersections. By assumption, $\mathscr{F} \subseteq \mathscr{G}$. Therefore $\mathscr{G}$ contains the smallest collection of subsets of $[0, 1)^d$ that contains $\mathscr{F}$ and is closed under monotone unions and intersections. Since this collection is a sigma-algebra by the monotone class theorem, $\mathscr{G}$ contains a sigma-algebra that contains $\mathscr{F}$. It is easy to see that the smallest sigma-algebra containing $\mathscr{F}$ is the Borel sigma-algebra of $[0, 1)^d$. Thus,

$$\int_B f(x)\, dx = 0$$

for every Borel subset $B \subseteq [0, 1)^d$. Take any $\epsilon > 0$ and let $B_\epsilon^+ := \{x \in [0, 1)^d : f(x) > \epsilon\}$. Then

$$0 = \int_{B_\epsilon^+} f(x)\, dx \ge \epsilon \operatorname{Leb}(B_\epsilon^+) \ge 0,$$

where $\operatorname{Leb}(B_\epsilon^+)$ is the Lebesgue measure if B_ϵ^+. This shows that $\operatorname{Leb}(B_\epsilon^+) = 0$. Similarly if $B_\epsilon^- := \{x \in [0, 1)^d : f(x) < -\epsilon\}$, then $\operatorname{Leb}(B_\epsilon^-) = 0$. This proves that $f = 0$ almost everywhere on $[0, 1]^d$. $\qquad\square$

Recall the definition of $\mathscr{D}_k$, the set of all closed dyadic cubes of side-length 2^{-k} in $[0, 1]^d$. Take any $f \in L^2([0, 1]^d)$ and an integer $k \ge 0$. Let f_k be the function that equals

$$\frac{1}{\operatorname{Leb}(D)} \int_D f(x)\, dx$$

in the interior of every $D \in \mathscr{D}_k$, and equals zero on the boundaries. We will refer to f_k as the 'level k dyadic approximant of f'.

Proposition 2.5 *Take any $f \in L^2([0, 1]^d)$ and let f_k be the level k dyadic approximant of f. Then f_k converges to f in L^2 as $k \to \infty$.*

Proof Take any $0 \leq k < l$, and a cube $D \in \mathcal{D}_k$. Let $\mathcal{C}$ be the set of all members of $\mathcal{D}_l$ that are contained in D. For each $C \in \mathcal{C}$, let x_C be the value of f_l in C. Similarly, let x_D be the value of f_k in D. Then note that

$$x_D = \frac{1}{|\mathcal{C}|} \sum_{C \in \mathcal{C}} x_C. \tag{2.2.1}$$

Therefore,

$$\frac{1}{\text{Leb}(D)} \int_D (f_l(x) - f_k(x))^2 \, dx = \frac{1}{|\mathcal{C}|} \sum_{C \in \mathcal{C}} (x_C - x_D)^2$$

$$= \left(\frac{1}{|\mathcal{C}|} \sum_{C \in \mathcal{C}} x_C^2 \right) - x_D^2$$

$$= \frac{1}{\text{Leb}(D)} \int_D (f_l(x)^2 - f_k(x)^2) \, dx.$$

Summing over all D, we get

$$\|f_l - f_k\|^2 = \|f_l\|^2 - \|f_k\|^2.$$

Using the Cauchy–Schwarz inequality, it is easy to see that for any k, $\|f_k\|^2 \leq \|f\|^2$. Combining this with the identity displayed above, it follows that $\{f_k\}_{k \geq 0}$ is a Cauchy sequence in $L^2([0, 1]^d)$. By the completeness of $L^2([0, 1]^d)$, there exists g such that $f_k \to g$ in L^2 as $k \to \infty$. Consequently, for any $D \in \mathcal{D}$,

$$\int_D g(x) \, dx = \lim_{k \to \infty} \int_D f_k(x) \, dx.$$

However, for any $D \in \mathcal{D}$,

$$\int_D f_k(x) \, dx = \int_D f(x) \, dx$$

for all large enough k. Thus, for every $D \in \mathcal{D}$,

$$\int_D (f(x) - g(x)) \, dx = 0.$$

By Proposition 2.4, this implies that $f = g$ almost everywhere.   □
For certain purposes, we will need a more general approximation result than Proposition 2.5. For a positive integer n, let $\mathcal{B}_n$ be the set of all cubes of the form

$$\left[\frac{i_1 - 1}{n}, \frac{i_1}{n} \right] \times \left[\frac{i_2 - 1}{n}, \frac{i_2}{n} \right] \times \cdots \times \left[\frac{i_d - 1}{n}, \frac{i_d}{n} \right].$$

Note that $\mathscr{D}_k = \mathscr{B}_{2^k}$. Given a function $f \in L^2([0,1]^d)$ and a positive integer n, define

$$\hat{f}_n(x) := \frac{1}{\text{Leb}(B)} \int_B f(x)\, dx$$

if x belongs to the interior of a cube $B \in \mathscr{B}_n$, and let $\hat{f}_n$ be zero on the boundaries of these cubes. Again, note that the dyadic approximant f_k is nothing but $\hat{f}_{2^k}$. We will call $\hat{f}_n$ the 'level n approximant' of f, dropping the word 'dyadic'. The following result generalizes Proposition 2.5 to general approximants.

Proposition 2.6 *Take any $f \in L^2([0,1]^d)$ and let $\hat{f}_n$ be the level n approximant of f, as defined above. Then $\hat{f}_n$ converges to f in L^2 as $n \to \infty$.*

Proof Fix some $k \geq 0$. Let f_k be the level k dyadic approximant of f. Take any $n \geq 2^k$. Classify the elements of $\mathscr{B}_n$ into two groups, as follows. Let $\mathscr{B}'_n$ be the set of all $B \in \mathscr{B}_n$ that are fully contained in some $D \in \mathscr{D}_k$. Let $\mathscr{B}''_n$ be the set of all elements of $\mathscr{B}_n$ that do not have the above property.

Take any $B \in \mathscr{B}'_n$. Then by the Cauchy–Schwarz inequality,

$$\int_B (\hat{f}_n(x) - f_k(x))^2\, dx = \int_B \left(\frac{1}{\text{Leb}(B)} \int_B (f(y) - f_k(y))\, dy \right)^2 dx$$

$$\leq \int_B (f(x) - f_k(x))^2\, dx.$$

Therefore

$$\sum_{B \in \mathscr{B}'_n} \int_B (\hat{f}_n(x) - f_k(x))^2\, dx \leq \|f - f_k\|^2. \tag{2.2.2}$$

Similarly for any $B \in \mathscr{B}''_n$,

$$\int_B (\hat{f}_n(x) - f_k(x))^2\, dx \leq 2 \int_B \hat{f}_n(x)^2\, dx + 2 \int_B f_k(x)^2\, dx$$

$$= 2 \int_B \left(\frac{1}{\text{Leb}(B)} \int_B f(y)\, dy \right)^2 dx + 2 \int_B f_k(x)^2\, dx$$

$$\leq 2 \int_B f(x)^2\, dx + 2 \int_B f_k(x)^2\, dx.$$

This shows that

$$\sum_{B \in \mathscr{B}''_n} \int_B (\hat{f}_n(x) - f_k(x))^2\, dx \leq 2 \int_{[0,1]^d} (f(x)^2 + f_k(x)^2) \psi_n(x)\, dx, \tag{2.2.3}$$

awhere

$$\psi_n(x) = \begin{cases} 1 & \text{if } x \in B \text{ for some } B \in \mathcal{B}_n'', \\ 0 & \text{otherwise.} \end{cases}$$

It is easy to see that $\psi_n(x) \to 0$ as $n \to \infty$ for almost every $x \in [0,1]^d$. Therefore by the dominated convergence theorem, the right-hand side of (2.2.3) tends to zero as $n \to \infty$. Combining this with (2.2.2) gives

$$\limsup_{n\to\infty} \|\hat{f}_n - f_k\| \le \|f - f_k\|,$$

and therefore,

$$\limsup_{n\to\infty} \|\hat{f}_n - f\| \le 2\|f - f_k\|.$$

Since this holds for every k, Proposition 2.5 implies that $\hat{f}_n$ converges to f in L^2. □
The following corollary of Proposition 2.6 will be useful later.

Corollary 2.1 *Let $f \in L^2([0,1]^d)$ and let $\hat{f}_n$ be the level n approximant of f, as defined above. Let $\phi : I \to \mathbb{R}$ be a bounded continuous function, where I is any interval containing the range of (some version of) f. Then*

$$\lim_{n\to\infty} \int_{[0,1]^d} \phi(\hat{f}_n(x))\,dx = \int_{[0,1]^d} \phi(f(x))\,dx.$$

Proof Let L_n denote the integral on the left-hand side and L be the integral on the right. Note that L_n is well-defined since I contains the range of $\hat{f}_n$ for any n. Let $\{n_k\}_{k\ge1}$ be a sequence of integers tending to infinity. Since $\hat{f}_{n_k}$ tends to f in L^2 by Proposition 2.6, Proposition 2.2 implies the existence of a further subsequence $\{n_{k_l}\}_{l\ge1}$ such that $\hat{f}_{n_{k_l}} \to f$ almost everywhere. By the dominated convergence theorem, this implies that $L_{n_{k_l}} \to L$ as $l \to \infty$. Thus, we have shown that for any sequence n_k, there is a subsequence n_{k_l} such that $L_{n_{k_l}} \to L$. This shows that $L_n \to L$. □

2.3 The Weak Topology and Its Compactness

There is a natural inner product on $L^2([0,1]^d)$, defined as

$$(f,g) := \int_{[0,1]^d} f(x)g(x)\,dx.$$

Note that for any $g \in L^2([0,1]^d)$, the map $f \mapsto (f,g)$ is continuous. The *weak topology* on $L^2([0,1]^d)$ is defined as the smallest topology under which the map $f \mapsto (f,g)$ is continuous for every $g \in L^2([0,1]^d)$.

Take any $g_1, \ldots, g_k \in L^2([0,1]^d)$, any $x_1, \ldots, x_k \in \mathbb{R}$ and $\epsilon_1, \ldots, \epsilon_k > 0$. Consider the set

$$V := \{ f \in L^2([0,1]^d) : |(f, g_i) - x_i| < \epsilon_i \text{ for } i = 1, \ldots, k\}. \tag{2.3.1}$$

Clearly, V is an open set in the weak topology. Consider the collection $\mathcal{T}$ of all possible unions of sets of the above form. Since sets like V are open in the weak topology, $\mathcal{T}$ is therefore a collection of open sets in the weak topology. (In topological terminology, a collection like V is called a basis for the topology $\mathcal{T}$.) Since sets like V are closed under finite intersections, $\mathcal{T}$ is itself a topology. Moreover, for any $g \in L^2([0,1]^d)$, the map $f \mapsto (f, g)$ is continuous in this topology. Therefore, the topology $\mathcal{T}$ must be the same as the weak topology. In other words, any open set in the weak topology is a union of sets like V.

Let $B_1([0,1]^d)$ denote the closed unit ball of $L^2([0,1]^d)$. The weak topology on this set is of particular importance in this monograph. We will now show that $B_1([0,1]^d)$ is metrizable and compact under the weak topology.

Recall the set $\mathcal{D}$ of closed dyadic cubes in $[0,1]^d$. Let $D_1, D_2, \ldots$ be an enumeration of the members of $\mathcal{D}$. For two functions $f, g \in B_1([0,1]^d)$, define

$$\delta(f,g) := \sum_{m=1}^{\infty} 2^{-m} \min\left\{ \left| \int_{D_m} (f(x) - g(x)) \, dx \right|, 1 \right\}.$$

It is easy to verify that δ is symmetric and satisfies the triangle inequality. By Proposition 2.4, $\delta(f,g) = 0$ if and only if $f = g$ almost everywhere. Therefore δ is a metric on $B_1([0,1]^d)$.

Proposition 2.7 *The metric δ defined above metrizes the restriction of the weak topology to $B_1([0,1]^d)$.*

Proof Take any $f \in B_1([0,1]^d)$ and any $\epsilon > 0$. Let M be so large that $2^{-M} < \epsilon/2$. Let

$$U := \left\{ h \in B_1([0,1]^d) : \left| \int_{D_m} (h(x) - f(x)) \, dx \right| < \epsilon/2 \text{ for } m = 1, \ldots, M \right\}.$$

Then U is an open set in the weak topology restricted to $B_1([0,1]^d)$, and for any $h \in U$,

$$\delta(h,f) < \sum_{m=1}^{M} 2^{-m} \frac{\epsilon}{2} + \sum_{m=M+1}^{\infty} 2^{-m}$$

$$\leq \frac{\epsilon}{2} + 2^{-M} \leq \epsilon.$$

This shows that any open ball of the metric δ is open in the weak topology in $B_1([0,1]^d)$.

Next, consider the set V defined in (2.3.1) and let $V' := V \cap B_1([0,1]^d)$. Take any $f \in V'$. Then there exist positive numbers $\kappa_1, \ldots, \kappa_k$ such that for each i,

$$|(f,g_i) - x_i| < \epsilon_i - \kappa_i.$$

For each i, let h_i be a dyadic approximant of g_i such that

$$\|g_i - h_i\| < \frac{\kappa_i}{3}.$$

By Proposition 2.5, such approximants exist. By the Cauchy–Schwarz inequality and the fact that $\|f\| \leq 1$, this implies that

$$|(f,g_i) - (f,h_i)| < \frac{\kappa_i}{3}.$$

Since the h_i's are constant on certain dyadic cubes, there exists a positive number ϵ so small that whenever $\delta(f,u) < \epsilon$, we have

$$|(f,h_i) - (u,h_i)| < \frac{\kappa_i}{3}$$

for each i. Consequently, if $\delta(f,u) < \epsilon$ and $\|u\| \leq 1$, then

$$|(u,g_i) - x_i| \leq |(u,g_i) - (u,h_i)| + |(u,h_i) - (f,h_i)|$$
$$+ |(f,h_i) - (f,g_i)| + |(f,g_i) - x_i|$$
$$< \|u\|\|g_i - h_i\| + \frac{\kappa_i}{3} + \frac{\kappa_i}{3} + \epsilon_i - \kappa_i \leq \epsilon_i.$$

In other words, the intersection of $B_1([0,1]^d)$ and the δ-ball of radius ϵ centered at f is contained in V'. This shows that any weakly open subset of $B_1([0,1]^d)$ is open in the topology defined by δ, completing the proof. $\square$

Proposition 2.8 (Special Case of the Banach–Alaoglu Theorem, Without Axiom of Choice) *The weak topology on $B_1([0,1]^d)$ is compact.*

Proof Since we have shown that the weak topology is metrizable, it is enough to show that any sequence has a convergent subsequence. Accordingly, let $\{f_n\}_{n \geq 1}$ be a sequence of functions in $B_1([0,1]^d)$. Let $D_1, D_2, \ldots$ be an enumeration of dyadic cubes, as before. Note that for any m and n,

$$\left| \int_{D_m} f_n(x)\, dx \right| \leq \|f_n\| \leq 1.$$

Therefore by a standard diagonal argument, there exists a subsequence $\{f_{n_k}\}_{k\geq 1}$ such that the limit

$$I(D_m) := \lim_{k\to\infty} \int_{D_m} f_{n_k}(x)\,dx$$

exists for each m.

Take any $k \geq 0$ and consider the set $\mathscr{D}_k$ of dyadic cubes at level k. Define a function g_k that is identically equal to $I(D)/\mathrm{Leb}(D)$ on each $D \in \mathscr{D}_k$. From the definition of $I(D)$ it is easy to verify that the g_k's have the same averaging property (2.2.1) as the f_k's in Proposition 2.5. As a consequence, just like the f_k's of Proposition 2.5, the g_k's also have the property that for any $0 \leq k < l$, $\|g_l - g_k\|^2 = \|g_l\|^2 - \|g_k\|^2$. Moreover, it follows from the definition of the $I(D)$'s that $\|g_k\| \leq 1$ for each k. Therefore the sequence $\{g_k\}_{k\geq 0}$ is Cauchy in L^2, and thus, converges to a limit function g. It is easy to see that this limit satisfies $\|g\| \leq 1$ and

$$\int_D g(x)\,dx = I(D)$$

for every $D \in \mathscr{D}$. Consequently, the sequence $\{f_{n_k}\}_{k\geq 1}$ converges to the function g in the δ metric. Therefore by Proposition 2.7, f_{n_k} converges weakly to g. □

2.4 Compact Operators

A linear operator K on $L^2([0,1]^d)$ is a linear map from $L^2([0,1]^d)$ into itself. The norm of K is defined as

$$\|K\| := \sup\{\|Kf\| : f \in L^2([0,1]^d),\ \|f\| = 1\}.$$

A linear operator is called bounded if its norm is finite, and self-adjoint if

$$(Kf, g) = (g, Kf)$$

for all $f, g \in L^2([0,1]^d)$. The following lemma gives a useful alternate expression for the norm of a bounded self-adjoint operator.

Lemma 2.3 *If K is a bounded self-adjoint linear operator, then*

$$\|K\| = \sup_{\|u\|=1} |(u, Ku)|.$$

Proof By the Cauchy–Schwarz inequality, for any u with $\|u\| = 1$,

$$|(u, Ku)| \leq \|u\|\|Ku\| \leq \|K\|.$$

Conversely, let $\lambda := \sup_{\|u\|=1} |(u, Ku)|$. Then for any u and v,

$$|(u + v, K(u + v)) - (u - v, K(u - v))| \leq \lambda \|u + v\|^2 + \lambda \|u - v\|^2.$$

This can be rewritten as

$$4|(u, Kv)| \leq 2\lambda(\|u\|^2 + \|v\|^2).$$

Now take any v with $\|v\| = 1$. We want to show that $\|Kv\| \leq \lambda$. If $Kv = 0$ then this is true anyway. So assume that $Kv \neq 0$. Let $u := Kv/\|Kv\|$. Then by the above inequality,

$$\|Kv\| = (u, Kv) \leq \frac{\lambda}{2}(\|u\|^2 + \|v\|^2) = \lambda.$$

This completes the proof of the lemma. $\square$

Definition 2.1 A linear operator K on $L^2([0, 1]^d)$ is called compact if the image of $B_1([0, 1]^d)$ under K is compact in the L^2 topology.
An immediate consequence of the definition is that any compact operator is bounded. The Banach–Alaoglu theorem gives the following useful criterion for checking whether an operator is compact.

Proposition 2.9 *If an operator K is continuous as a map from $B_1([0, 1]^d)$ with the weak topology into $L^2([0, 1]^d)$ with the L^2 topology, then K is compact.*

Proof By Proposition 2.8, $B_1([0, 1]^d)$ is compact under the weak topology. Since the image of a compact set under a continuous map is compact, this completes the proof. $\square$
The following result establishes the existence of at least one eigenvalue for self-adjoint compact operators.

Proposition 2.10 *Let K be a self-adjoint compact linear operator on $L^2([0, 1]^d)$. Then there exists $u \in B_1([0, 1]^d)$ and a real number λ such that $|\lambda| = \|K\|$ and $Ku = \lambda u$.*

Proof If $K = 0$ then the proof is trivial. So let us assume that $\|K\| > 0$. By Lemma 2.3, there exists a sequence $\{u_n\}_{n \geq 1}$ in $B_1([0, 1]^d)$ such that

$$\lim_{n \to \infty} |(u_n, Ku_n)| = \|K\|.$$

Passing to a subsequence if necessary, we may assume that

$$\lambda := \lim_{n \to \infty} (u_n, Ku_n)$$

exists. By the definition of a compact operator and Proposition 2.8, we may pass to a further subsequence and assume that there exists $u \in B_1([0, 1]^d)$ such that

$Ku_n \to Ku$ in the L^2 topology and $u_n \to u$ in the weak topology. Consequently, $(u_n, Ku) \to (u, Ku)$ and

$$|(u_n, Ku_n) - (u_n, Ku)| \le \|u_n\| \|Ku_n - Ku\| \le \|Ku_n - Ku\| \to 0.$$

Therefore

$$(u, Ku) = \lambda.$$

Note that $\|u\|$ must be equal to 1, since clearly $u \ne 0$, and if $0 < \|u\| < 1$, then there exists $\alpha > 1$ such that $\|\alpha u\| = 1$, and $|(\alpha u, K(\alpha u))| = \alpha^2 |\lambda| > \|K\|$, which is impossible.

Take any $v \in L^2([0,1])$ and $\delta > 0$. Suppose that $\lambda > 0$. Then by the definition of $\|K\|$,

$$(u + \delta v, K(u + \delta v)) \le \lambda \|u + \delta v\|^2.$$

Using the fact that K is self-adjoint, this can be rewritten as

$$(u, Ku) + 2\delta(v, Ku) + \delta^2(v, Kv) \le \lambda \|u\|^2 + 2\lambda\delta(v, u) + \lambda\delta^2 \|v\|^2,$$

which is the same as

$$2\delta(v, Ku - \lambda u) \le \lambda \|u\|^2 - (u, Ku) + \lambda\delta^2 \|v\|^2 - \delta^2(v, Kv)$$
$$= \lambda\delta^2 \|v\|^2 - \delta^2(v, Kv).$$

Dividing both sides by δ and letting $\delta \to 0$, we get

$$(v, Ku - \lambda u) \le 0.$$

Taking $v = Ku - \lambda u$ in the above inequality shows that $Ku = \lambda u$, completing the proof in the case $\lambda > 0$. If $\lambda < 0$, the proof is completed in a similar manner starting from the inequality $(u + \delta v, K(u + \delta v)) \ge \lambda \|u + \delta v\|^2$ and choosing $v = -(Ku - \lambda u)$. □

We will say that a function $u \in B_1([0,1]^d)$ is nonnegative, and write $u \ge 0$, if $u \ge 0$ almost everywhere. More generally, $u \le v$ will mean that $u \le v$ almost everywhere. We will say that a linear operator K on $L^2([0,1]^d)$ is nonnegative, and write $K \ge 0$, if $Ku \ge 0$ whenever $u \ge 0$. The following result is an improvement of Proposition 2.10 for nonnegative operators.

Proposition 2.11 (Existence of Perron–Frobenius Eigenvalue for Nonnegative Operators) *If K is a self-adjoint compact nonnegative linear operator on $L^2([0,1]^d)$, then there exists $u \in B_1([0,1]^d)$ that is nonnegative everywhere, and satisfies $Ku = \|K\|u$.*

Proof If $K = 0$ then there is nothing to prove, so let us assume that $\|K\| > 0$. Let S be the set of nonnegative elements of $B_1([0, 1]^d)$. For each $u \in S$, let

$$L(u) := \sup\{\lambda \in \mathbb{R} : \lambda u \le Ku\}.$$

Since $K \ge 0$, $L(u) \ge 0$ for each $u \in S$. Define

$$L^* := \sup_{u \in S} L(u).$$

It is easy to see that $L(u) \le \|K\|$ for all $u \in S$, and therefore $L^* \le \|K\|$.

For each $v \in L^2([0, 1]^d)$, let $|v|$ denote the function obtained by taking the absolute value of $v(x)$ at every x. Note that $|v| \ge v$ and $|v| \ge -v$. Since K is a nonnegative operator, this implies that $K|v| \ge Kv$ and $K|v| \ge -Kv$. Thus,

$$K|v| \ge |Kv|.$$

By Proposition 2.10, there exists $v \in B_1([0, 1]^d)$ and $\lambda \in \mathbb{R}$ such that $Kv = \lambda v$ and $|\lambda| = \|K\|$. Let $u := |v|$. Then $u \in S$, and by the above inequality,

$$\|K\|u = |\lambda v| = |Kv| \le K|v| = Ku.$$

Thus, $L^* \ge L(u) \ge \|K\|$. Combining this with our previous observation that $L^* \le \|K\|$, we get

$$L^* = \|K\|. \qquad (2.4.1)$$

Let $\{u_n\}_{n \ge 1}$ be a sequence in S such that $L(u_n) \to L^*$. Then there is a sequence of nonnegative numbers $\{L_n\}_{n \ge 1}$ such that $L_n \to L^*$ and $L_n u_n \le K u_n$ for each n. Let $v_n := K u_n$. Since K is a compact operator, we may assume without loss of generality that v_n converges to some v in L^2. Since $K u_n \ge L_n u_n \ge 0$,

$$\|v\| = \lim_{n \to \infty} \|v_n\|$$

$$= \lim_{n \to \infty} \|K u_n\| \ge \lim_{n \to \infty} L_n \|u_n\| = L^*.$$

In particular, by (2.4.1),

$$\|v\| \ge \|K\| > 0. \qquad (2.4.2)$$

Next, note that since $K \ge 0$ and $L_n u_n \le K u_n$,

$$L_n v_n = K(L_n u_n) \le K(K u_n) = K v_n.$$

Since K is continuous and $v_n \to v$ in L^2, this gives

$$L^* v \leq Kv. \tag{2.4.3}$$

Lastly, observe that since $v_n \geq 0$ for every n, Proposition 2.2 implies that

$$v \geq 0. \tag{2.4.4}$$

By (2.4.2), we are allowed to define $w := v/\|v\|$. By (2.4.4), $w \geq 0$, and by (2.4.3), $L^* w \leq Kw$. However, this means that $L^* w$ must be equal to Kw, since otherwise, the nonnegativity of w and the inequality $L^* w \leq Kw$ would imply that

$$L^* = \|L^* w\| < \|Kw\| \leq \|K\|,$$

contradicting (2.4.1). By (2.4.1), this completes the proof. $\qquad\square$

2.5 A Generalized Hölder's Inequality

The following non-trivial generalization of Hölder's inequality will be useful later for our analysis of large deviations for random graphs.

Theorem 2.1 *Let $\mu_1, \ldots, \mu_n$ be probability measures on $\Omega_1, \ldots, \Omega_n$, respectively, and let $\mu = \prod_{i=1}^{n} \mu_i$ be the product measure on $\Omega = \prod_{i=1}^{n} \Omega_i$. Let $A_1, \ldots, A_m$ be non-empty subsets of $[n] = \{1, 2, \ldots, n\}$ and write $\Omega_A = \prod_{l \in A} \Omega_l$ and $\mu_A = \prod_{l \in A} \mu_l$. Let $f_i \in L^{p_i}(\Omega_{A_i}, \mu_{A_i})$ with $p_i \geq 1$ for each $i \in [m]$ and suppose that for each $l \in [n]$,*

$$\sum_{i: l \in A_i} \frac{1}{p_i} \leq 1.$$

Each f_i can be thought of as an element of $L^{p_i}(\Omega, \mu)$ in a natural way. With this interpretation, the following inequality holds:

$$\int_{\Omega} \prod_{i=1}^{m} |f_i| \, d\mu \leq \prod_{i=1}^{m} \left(\int_{\Omega_{A_i}} |f_i|^{p_i} \, d\mu_{A_i} \right)^{1/p_i}.$$

In particular, when each $l \in [n]$ belongs to at most d many A_i's, then we can take $p_i = d$ for every $i \in [m]$ and get

$$\int_{\Omega} \prod_{i=1}^{m} |f_i| \, d\mu \leq \prod_{i=1}^{m} \left(\int_{\Omega_{A_i}} |f_i|^{d} \, d\mu_{A_i} \right)^{1/d}.$$

Proof The proof is by induction on n. The case $n = 1$ is ordinary Hölder's inequality. Suppose that $n > 1$ and that the inequality holds for all smaller values of n. By Fubini's theorem,

$$\int_\Omega \prod_{i=1}^m |f_i| \, d\mu = \int_\Omega \prod_{i:n\in A_i} |f_i| \prod_{i:n\notin A_i} |f_i| \, d\mu$$

$$= \int_{\Omega_{[n-1]}} \left(\int_{\Omega_n} \prod_{i:n\in A_i} |f_i| \, d\mu_n \right) \prod_{i:n\notin A_i} |f_i| \, d\mu_{[n-1]}.$$

For each i such that $n \in A_i$, define a function $f_i^* : \Omega_{[n-1]} \to \mathbb{R}$ as

$$f_i^* := \left(\int_{\Omega_n} |f_i|^{p_i} \, d\mu_n \right)^{1/p_i}.$$

By Hölder's inequality,

$$\int_{\Omega_n} \prod_{i:n\in A_i} |f_i| \, d\mu_n \le \prod_{i:n\in A_i} \left(\int_{\Omega_n} |f_i|^{p_i} \, d\mu_n \right)^{1/p_i} = \prod_{i:n\in A_i} f_i^*.$$

Substituting this in the identity obtained above, we get

$$\int_\Omega \prod_{i=1}^m |f_i| \, d\mu \le \int_{\Omega_{[n-1]}} \prod_{i:n\in A_i} f_i^* \prod_{i:n\notin A_i} |f_i| \, d\mu_{[n-1]}.$$

Applying the induction hypothesis for $n - 1$ to the right side gives

$$\int_{\Omega_{[n-1]}} \prod_{i:n\in A_i} f_i^* \prod_{i:n\notin A_i} |f_i| \, d\mu_{[n-1]}$$

$$\le \prod_{i:n\in A_i} \left(\int_{\Omega_{[n-1]}} (f_i^*)^{p_i} \, d\mu_{[n-1]} \right)^{1/p_i} \prod_{i:n\notin A_i} \left(\int_{\Omega_{[n-1]}} |f_i|^{p_i} \, d\mu_{[n-1]} \right)^{1/p_i}$$

$$= \prod_{i=1}^m \left(\int_\Omega |f_i|^{p_i} \, d\mu \right)^{1/p_i},$$

as required. $\square$

2.6 The FKG Inequality

Let S be a finite or countable subset of $\mathbb{R}$. Let n be a positive integer, and let ρ be a probability mass function on S^n. If $x = (x_1, \ldots, x_n)$ and $y = (y_1, \ldots, y_n)$ are two elements of S^n, we will write $x \leq y$ if $x_i \leq y_i$ for each i. A function $f : S^n \to \mathbb{R}$ is called monotone increasing if $f(x) \leq f(y)$ whenever $x \leq y$. For $x, y \in S^n$, $x \vee y$ denotes the vector whose ith coordinate is the maximum of x_i and y_i. Similarly, $x \wedge y$ denotes the vector whose ith coordinate is the minimum of x_i and y_i. The probability mass function ρ is said to satisfy the FKG lattice condition if for all $x, y \in S^n$,

$$\rho(x)\rho(y) \leq \rho(x \vee y)\rho(x \wedge y). \tag{2.6.1}$$

Recall that the covariance of two random variables X and Y is defined as

$$\mathrm{Cov}(X, Y) := \mathbb{E}(XY) - \mathbb{E}(X)\mathbb{E}(Y).$$

The FKG inequality says the following.

Theorem 2.2 (FKG Inequality) *Let S, n and ρ be as above and suppose that ρ satisfies (2.6.1) and is strictly positive everywhere on S^n. Let X be a random vector with law ρ. Then for any monotone increasing $f, g : S^n \to \mathbb{R}$ such that $f(X)$ and $g(X)$ are square-integrable random variables, $\mathrm{Cov}(f(X), g(X)) \geq 0$.*

Proof The proof is by induction on n. First, suppose that $n = 1$. Let X, Y be independent random variables with law ρ. Then by monotonicity of f and g, $(f(X) - f(Y))(g(X) - g(Y))$ is a nonnegative random variable. By the independence of X and Y, it follows that

$$\mathrm{Cov}(f(X), g(X)) = \mathbb{E}(f(X)g(X)) - \mathbb{E}(f(X))\mathbb{E}(g(X))$$

$$= \frac{1}{2}\mathbb{E}[(f(X) - f(Y))(g(X) - g(Y))].$$

This proves the claim when $n = 1$. Next, suppose that $n > 1$ and that the theorem has been proved in all smaller dimensions. Define, for each $a \in S$,

$$f_1(a) := \mathbb{E}(f(X) \mid X_1 = a), \quad g_1(a) := \mathbb{E}(g(X) \mid X_1 = a).$$

Then by a well-known and easy-to-prove identity about covariances,

$$\mathrm{Cov}(f(X), g(X))$$
$$= \mathbb{E}(\mathrm{Cov}(f(X), g(X) \mid X_1)) + \mathrm{Cov}(\mathbb{E}(f(X) \mid X_1), \mathbb{E}(g(X) \mid X_1))$$
$$= \mathbb{E}(\mathrm{Cov}(f(X), g(X) \mid X_1)) + \mathrm{Cov}(f_1(X_1), g_1(X_1)).$$

If the first coordinate is fixed, then f and g are monotone increasing functions of the remaining $n - 1$ coordinates. Also, it is easy to see that the condition probability mass function of $(X_2, \ldots, X_n)$ given $X_1 = a$ satisfies the lattice condition (2.6.1) on S^{n-1}, for any a. Combining these two observations, it follows from the induction hypothesis that for any $a \in S$,

$$\mathrm{Cov}(f(X), g(X) \mid X_1 = a) \geq 0.$$

Thus, the proof will be complete if we can show that f_1 and g_1 are monotone increasing functions on S. By symmetry, it suffices to show only for f_1. Take any $a, b \in S, a < b$. Let $X' := (X_2, \ldots, X_n)$ and for each $x' \in S^{n-1}$, let

$$\tau(x') := \frac{\rho(b, x')}{\rho(a, x')}.$$

Then by the monotonicity of f,

$$
\begin{aligned}
f_1(b) - f_1(a) &= \frac{\sum_{x' \in S^{n-1}} f(b, x') \rho(b, x')}{\sum_{x' \in S^{n-1}} \rho(b, x')} - f_1(a) \\
&\geq \frac{\sum_{x' \in S^{n-1}} f(a, x') \rho(b, x')}{\sum_{x' \in S^{n-1}} \rho(b, x')} - f_1(a) \\
&= \frac{\mathbb{E}(f(a, X') \tau(X') \mid X_1 = a)}{\mathbb{E}(\tau(X') \mid X_1 = a)} - \mathbb{E}(f(a, X') \mid X_1 = a) \\
&= \frac{\mathrm{Cov}(f(a, X'), \tau(X') \mid X_1 = a)}{\mathbb{E}(\tau(X') \mid X_1 = a)}.
\end{aligned}
$$

Now $f(a, \cdot)$ is a monotone increasing function of S^{n-1}, and as observed before, the conditional law of X' given $X_1 = a$ satisfies (2.6.1). Thus, the above display shows that the proof of the monotonicity of f_1 will be complete if we can prove that τ is a monotone increasing function on S^{n-1}. To prove this, take any $x', y' \in S^{n-1}, x' \leq y'$. Then

$$\tau(y') - \tau(x') = \frac{\rho(b, y') \rho(a, x') - \rho(b, x') \rho(a, y')}{\rho(a, y') \rho(a, x')}.$$

But $\rho(b, y') \rho(a, x') - \rho(b, x') \rho(a, y') \geq 0$ by (2.6.1). This completes the proof of the theorem. $\square$

Bibliographical Notes

Most of the topics covered in this chapter are standard fare in graduate-level functional analysis and probability. I chose to restrict attention to $L^2([0, 1]^d)$ instead of general L^2 spaces because this is all that we need in this monograph. The specialization to the hypercube allows shorter proofs for several theorems. Further discussions and applications of Chebychev's inequality, Borel–Cantelli lemma and the monotone class theorem may be found in any graduate text on probability theory. Similarly, discussions on completeness of L^2, weak topology, the Banach–Alaoglu theorem and compact operators may be found in any graduate text on functional analysis. The discrete approximation presented in Sect. 2.2 is harder to find in textbooks; it is, however, very important for this monograph.

The generalized Hölder's inequality in Sect. 2.5 is less standard than the rest of the chapter. It is due to Finner [1], and also appears in Friedgut [3]. The statement and proof given here follow the presentation in Lubetzky and Zhao [4].

The FKG inequality is due to Fortuin et al. [2]. Many sophisticated generalized versions are now available, but the version stated and proved in Sect. 2.6 is the one we need in this monograph.

References

1. Finner, H. (1992). A generalization of Hölder's inequality and some probability inequalities. *The Annals of Probability, 20*(4), 1893–1901.
2. Fortuin, C. M., Kasteleyn, P. W., & Ginibre, J. (1971). Correlation inequalities on some partially ordered sets. *Communications in Mathematical Physics, 22*, 89–103.
3. Friedgut, E. (2004). Hypergraphs, entropy, and inequalities. *The American Mathematical Monthly, 111*(9), 749–760.
4. Lubetzky, E., & Zhao, Y. (2015). On replica symmetry of large deviations in random graphs. *Random Structures Algorithms, 47*(1), 109–146.

Chapter 3
Basics of Graph Limit Theory

This chapter summarizes some basic results from graph limit theory. The only background assumed here is the list of results from Chap. 2. As in Chap. 2, I will try to make the presentation and the proofs as self-contained as possible.

3.1 Graphons and Homomorphism Densities

A simple graph is an undirected graph with no multi-edges or self-loops. Let G be a finite simple graph on n vertices. Let $V(G) = \{1, 2, \ldots, n\}$ be the set of vertices of G and let $E(G)$ be the set of edges. Recall that the adjacency matrix of G is the $n \times n$ symmetric matrix whose (i, j)th entry is 1 if $\{i, j\} \in E(G)$ and 0 if not. The adjacency matrix may be converted into a function f^G on $[0, 1]^2$ in the following canonical way: Take any $(x, y) \in [0, 1]^2$, and let i and j be the two unique integers such that

$$\frac{i-1}{n} < x \le \frac{i}{n} \quad \text{and} \quad \frac{j-1}{n} < y \le \frac{j}{n}.$$

Then define

$$f^G(x, y) = \begin{cases} 1 & \text{if } \{i, j\} \in E(G), \\ 0 & \text{if not.} \end{cases}$$

The function f^G is called the graphon of G. Notice that f^G is a Borel measurable function and $f^G(x, y) = f^G(y, x)$. These two properties characterize the general definition of a graphon given below. Graphons of finite graphs are special examples of this general definition.

© Springer International Publishing AG 2017
S. Chatterjee, *Large Deviations for Random Graphs*, Lecture Notes in Mathematics 2197, DOI 10.1007/978-3-319-65816-2_3

Definition 3.1 Let $\mathscr{W}$ be the set of all $f : [0, 1]^2 \to [0, 1]$ that are Borel measurable and satisfy $f(x, y) = f(y, x)$, after taking the quotient with respect to the equivalence relation of almost everywhere equality. An element of $\mathscr{W}$ is called a graphon.

Let H and G be two finite simple graphs. A homomorphism from H into G is a map $\varphi : V(H) \to V(G)$ that preserves edges. That is, whenever $\{v, w\} \in E(H)$, we need to have $\{\varphi(v), \varphi(w)\} \in E(G)$.

Let $\mathrm{hom}(H, G)$ denote the number of homomorphisms of H into G. For example, if H is a triangle, then $\mathrm{hom}(H, G)$ is the number of triangles in G multiplied by six, since each triangle in G yields six different homomorphisms of H into G. Notice that the maximum possible number of homomorphisms of a graph on k vertices into a graph on n vertices in n^k. This allows us to define a quantity called homomorphism density that is guaranteed to belong to the interval $[0, 1]$.

Definition 3.2 Let H and G be finite simple graphs. The homomorphism density of H in G is defined as

$$t(H, G) := \frac{\mathrm{hom}(H, G)}{|V(G)|^{|V(H)|}},$$

where $\mathrm{hom}(H, G)$ is the number of homomorphisms of H into G, as defined above. Given an arbitrary map $\varphi : V(H) \to V(G)$, it is a homomorphism if and only if $\{\varphi(i), \varphi(j)\} \in E(G)$ for every $\{i, j\} \in E(H)$. Using this, it is not difficult to prove the following.

Exercise 3.1 Suppose that $V(H) = \{1, 2, \ldots, k\}$. In terms of the graphon f^G, show that the homomorphism density of H in G may be expressed as

$$t(H, G) = \int_{[0,1]^k} \prod_{\{i,j\} \in E(H)} f^G(x_i, x_j) \, dx_1 \, dx_2 \cdots dx_k.$$

The above expression leads to the following definition of homomorphism density of H in a general graphon.

Definition 3.3 Let H be a finite simple graph and let $f \in \mathscr{W}$ be a graphon. The homomorphism density of H in f is defined as

$$t(H, f) := \int_{[0,1]^k} \prod_{\{i,j\} \in E(H)} f(x_i, x_j) \, dx_1 \, dx_2 \cdots dx_k.$$

Notice that with this definition, Exercise 3.1 implies that $t(H, G)$ is the same as $t(H, f^G)$. Thus, the set of all finite simple graphs is a subset of the set of graphons. The following proposition shows that the set of graphons may be viewed as a compactification of the set of all finite simple graphs. This notion is made more precise in the subsequent sections.

Proposition 3.1 *For any graphon f, there exists a sequence of finite simple graphs $\{G_n\}_{n\geq 1}$ such that $t(H, G_n) \to t(H, f)$ for every H.*

Proof Let $\hat{f}_n$ be the level n approximant of f, as defined in Sect. 2.2 of Chap. 2. Let G_n be the random graph on n vertices with independent edges, where the edge $\{i, j\}$ is present with probability p_{ij}, where p_{ij} is the value of $\hat{f}_n$ in the box $((i-1)/n, i/n) \times ((j-1)/n, j/n)$. Take any finite simple graph H. It is easy to check that

$$\lim_{n \to \infty} (\mathbb{E}(t(H, G_n)) - t(H\hat{f}_n)) = 0.$$

Another easy calculation shows that

$$\lim_{n \to \infty} \text{Var}(t(H, G_n)) = 0.$$

By Proposition 2.6,

$$\lim_{n \to \infty} t(H\hat{f}_n) = t(H, f).$$

Combining all this, we see that $t(H, G_n) \to t(H, f)$ in L^2. Suppose that these random graphs are all defined on the same probability space. Then by Proposition 2.2, there is a subsequence along which $t(H, G_n)$ converges to $t(H, f)$ almost surely. Since the set of finite simple graphs is countable, we can now use a diagonal argument to extract a subsequence along which $t(H, G_n) \to t(H, f)$ almost surely for every H. This completes the proof of the proposition. $\qquad\square$

3.2 The Cut Metric

The space $\mathscr{W}$ of graphons is equipped with a metric called the cut metric. It is defined as follows.

Definition 3.4 If $f, g \in \mathscr{W}$ are two graphons, the cut distance between f and g is defined as

$$d_\square(f, g) := \sup_{a,b} \left| \int_{[0,1]^2} a(x)b(y)(f(x, y) - g(x, y)) \, dx \, dy \right|,$$

where the supremum is taken over all Borel measurable $a, b : [0, 1] \to [-1, 1]$.

Exercise 3.2 Verify that $d_\square$ is indeed a metric on $\mathscr{W}$.

The following result shows that homomorphism densities are continuous with respect to the cut metric.

Proposition 3.2 (Counting Lemma) *For any finite simple graph H and any $f, g \in \mathscr{W}$,*

$$|t(H, f) - t(H, g)| \leq |E(H)| d_\square(f, g),$$

where $E(H)$ is the set of edges of H. In particular, the map $f \mapsto t(H, f)$ is a continuous function from $(\mathcal{W}, d_\square)$ into $\mathbb{R}$.

Proof Let f and g be two graphons. Let $V(H) = \{1, \dots, k\}$. Choose a distinguished edge $e \in E(H)$. Without loss of generality, suppose that $e = \{1, 2\}$. Let $E' := E(H) \setminus \{e\}$. Fix some values of $x_3, \dots, x_k$, and define two functions a and b and a constant c as

$$a(x) := \prod_{i:\{1,i\}\in E'} f(x, x_i),$$

$$b(x) := \prod_{i:\{2,i\}\in E'} f(x, x_i),$$

$$c := \prod_{\substack{i,j:\{i,j\}\in E' \\ i\geq 3, j\geq 3}} f(x_i, x_j).$$

Then notice that

$$\int_{[0,1]^2} (f(x_1, x_2) - g(x_1, x_2)) \prod_{\{i,j\}\in E'} f(x_i, x_j)\, dx_1\, dx_2$$

$$= c \int_{[0,1]^2} (f(x_1, x_2) - g(x_1, x_2)) a(x_1) b(x_2)\, dx_1\, dx_2.$$

Note that c, $a(x_1)$ and $b(x_2)$ are all bounded between 0 and 1. By the definition of the cut distance, this implies that if one $f(x_i, x_j)$ in the product within the integral in Definition 3.3 is replaced by $g(x_i, x_j)$, we will incur an error of at most $d_\square(f, g)$. This step can be repeated until every $f(x_i, x_j)$ has been replaced by the corresponding $g(x_i, x_j)$, yielding the bound

$$|t(H, f) - t(H, g)| \leq |E(H)| d_\square(f, g),$$

which completes the proof.   □

Note that $\mathcal{W}$ is a subset of $B_1([0, 1]^2)$, where $B_1([0, 1]^2)$ is the unit ball of $L^2([0, 1]^2)$, as defined in Chap. 2. As a subset of $B_1([0, 1]^2)$, $\mathcal{W}$ inherits the weak topology on $B_1([0, 1]^2)$ that was defined in Sect. 2.3 of Chap. 2. The following result gives the relation between the weak topology and the topology induced by $d_\square$. This has an important consequence in the next chapter.

Proposition 3.3 *The weak topology on $\mathcal{W}$ is weaker than the topology induced by the cut metric $d_\square$.*

Proof Recall the metric δ for the weak topology that was introduced in Sect. 2.3 of Chap. 2, and the sets $D_1, D_2, \dots$. To prove the claim, it suffices to show that whenever $d_\square(f_n, f) \to 0$, $\delta(f_n, f)$ also tends to zero. To show that $\delta(f_n, f) \to 0$, it suffices to prove that for each m,

$$\lim_{n \to \infty} \int_{D_m} (f_n(x, y) - f(x, y))\, dx\, dy = 0.$$

Since D_m is a square, the above integral can be written in the form

$$\int_{[0,1]^2} a(x)b(y)(f_n(x, y) - f(x, y))\, dx\, dy,$$

where a and b are Borel measurable functions from $[0, 1]$ into $[-1, 1]$. Since $d_\square(f_n, f) \to 0$, this integral must converge to zero as $n \to \infty$. $\quad\square$

3.3 Equivalence Classes of Graphons

A measure-preserving bijection of $[0, 1]$ is a map $\sigma : [0, 1] \to [0, 1]$ such that σ is a bijection, σ and σ^{-1} are Borel measurable, and for any Borel set A, the Lebesgue measures of $\sigma(A)$ and $\sigma^{-1}(A)$ are both equal to that of A. Let $\mathscr{M}$ be the set of all measure-preserving bijections of $[0, 1]$. If $\sigma \in \mathscr{M}$ and $f \in \mathscr{W}$, let

$$f_\sigma(x, y) := f(\sigma(x), \sigma(y)).$$

First proving for simple functions and then passing to the limit, it is easy to establish the following.

Exercise 3.3 Show that for any $f \in \mathscr{W}$ and $\sigma \in \mathscr{M}$,

$$\int_{[0,1]^2} f(x, y)\, dx\, dy = \int_{[0,1]^2} f_\sigma(x, y)\, dx\, dy.$$

From this it follows that the action of measure-preserving bijections leaves homomorphism densities invariant.

Exercise 3.4 Let f be a graphon, H be a finite simple graph, and σ be a measure-preserving bijection of $[0, 1]$. Prove that $t(H, f) = t(H, f_\sigma)$.

The metric $d_\square$ induces a pseudometric $\delta_\square$, defined as

$$\delta_\square(f, g) := \inf_{\sigma_1, \sigma_2} d_\square(f_{\sigma_1}, g_{\sigma_2}),$$

where the infimum is taken over all measure-preserving bijections σ_1 and σ_2. Notice that by Exercise 3.3, $\delta_\square$ may also be expressed as

$$\delta_\square(f, g) := \inf_\sigma d_\square(f_\sigma, g) = \inf_\sigma d_\square(f, g_\sigma).$$

The pseudometric naturally defines an equivalence relation, that leads to the definition of the following quotient space.

Definition 3.5 Declare that two graphons f and g equivalent if $\delta_\square(f, g) = 0$. Let $\widetilde{\mathcal{W}}$ be the set of all equivalence classes for this equivalence relation. For any $f \in \mathcal{W}$, let $\tilde{f}$ denote the equivalence class containing f. For any $f, g \in \mathcal{W}$, let $\delta_\square(\tilde{f}, \tilde{g}) := \delta_\square(f, g)$.

Exercise 3.5 Verify that $\delta_\square$ is well-defined on $\widetilde{\mathcal{W}}$ and is a metric.

The metric space $(\widetilde{\mathcal{W}}, \delta_\square)$ is particularly important in this monograph. Notice that by Proposition 3.2 and Exercise 3.4, we can define the homomorphism density of a graph H in an element of $\widetilde{\mathcal{W}}$, simply by letting $t(H, \tilde{f}) := t(H, f)$. Then the mapping $\tilde{f} \mapsto t(H, \tilde{f})$ is well-defined and continuous on $(\widetilde{\mathcal{W}}, \delta_\square)$. This is not true, for example, in the weak topology. The topology induced by $\delta_\square$ on $\widetilde{\mathcal{W}}$, as well as the topology induced by $d_\square$ on $\mathcal{W}$, will sometimes be called the cut topology in this manuscript.

3.4 Graphons as Operators

Any $f \in L^2([0, 1]^2)$ induces a linear operator K_f acting on $L^2([0, 1])$ as follows:

$$K_f u(x) = \int_0^1 f(x, y) u(y) \, dy. \tag{3.4.1}$$

A simple application of the Cauchy–Schwarz inequality shows that

$$\|K_f\| \le \|f\|, \tag{3.4.2}$$

where the norm on the left is operator norm and the norm on the right is L^2 norm. The following lemma connects the operator norm with the cut metric.

Lemma 3.1 *For any $f, g \in \mathcal{W}$, $d_\square(f, g) \le \|K_f - K_g\|$.*

Proof For any Borel measurable $a, b : [0, 1] \to [-1, 1]$, the Cauchy–Schwarz inequality and the definition of the operator norm imply that

$$\left| \int_{[0,1]^2} a(x) b(y) (f(x, y) - g(x, y)) \, dx \, dy \right| \le \|a\| \|K_{f-g} b\|$$

$$\le \|K_{f-g}\| = \|K_f - K_g\|.$$

Taking supremum over a and b completes the proof of the lemma. $\square$

The next result contains another important piece of information.

Proposition 3.4 *Let $f : [0, 1]^2 \to \mathbb{R}$ be bounded and Borel measurable. Then the operator K_f defined in (3.4.1) is compact.*

Proof By Propositions 2.7 and 2.9, it suffices to show that for any sequence $\{u_n\}_{n\geq 1}$ in $B_1([0, 1])$ that converges weakly to some u, $K_f u_n$ converges to $K_f u$ in L^2. Accordingly, take any such sequence. By the boundedness of f and the definition of the weak topology,

$$\lim_{n\to\infty} K_f u_n(x) = K_f u(x)$$

for every $x \in [0, 1]$. Next, note that if C is a constant such that $|f(x, y)| \leq C$ for all x and y, then by the Cauchy–Schwarz inequality,

$$|K_f u_n(x)| \leq C\|u_n\| \leq C$$

for all n and x. Therefore by the dominated convergence theorem, $K_f u_n \to K_f u$ in L^2. □

Recall that a function $u \in L^2([0, 1])$ is an eigenfunction of an operator K if $u \neq 0$ and $Ku = \lambda u$ for some $\lambda \in \mathbb{R}$. We will say that u is normalized if $\|u\| = 1$. A collection of eigenfunctions of an operator K will be called orthonormal if all the eigenfunctions are normalized, and the inner product between any two is zero.

We will now state and prove two technical results about eigenvalues and eigenfunctions of graphon-induced operators that will be needed in the next section.

Proposition 3.5 *Take any $f \in L^2([0, 1]^2)$ and $\epsilon > 0$. Let $\mathscr{C}$ be any orthonormal collection of eigenfunctions of the operator K_f such that none of the eigenvalues belong to the interval $(-\epsilon, \epsilon)$. Then $|\mathscr{C}| \leq \|f\|^2/\epsilon^2$. Moreover, if $|f(x, y)| \leq C$ everywhere, then for each $u \in \mathscr{C}$ and each x, $|u(x)| \leq C/\epsilon$.*

Proof For any $u \in \mathscr{C}$, let λ_u denote the corresponding eigenvalue, and let

$$g_u(x, y) := u(x)u(y).$$

Then note that $(g_u, g_v) = 0$ and $\|g_u\|^2 = 1$ whenever $u, v \in \mathscr{C}$ and $u \neq v$. Moreover,

$$(f, g_u) = \int_{[0,1]^2} f(x, y)u(x)u(y)\, dx\, dy = \int_{[0,1]} \lambda_u u(x)^2\, dx = \lambda_u.$$

Using the above identities it is easy to show that

$$\left\| f - \sum_{u\in\mathscr{C}} \lambda_u g_u \right\|^2 = \|f\|^2 - \sum_{u\in\mathscr{C}} \lambda_u^2,$$

which implies that

$$\epsilon^2 |\mathscr{C}| \leq \sum_{u \in \mathscr{C}} \lambda_u^2 \leq \|f\|^2.$$

This completes the proof of the first assertion. For the second, simply note that $|u(x)| = |\lambda_u^{-1} K_f u(x)| \leq \epsilon^{-1} |K_f u(x)|$ and use the Cauchy–Schwarz inequality to prove that $|K_f u(x)| \leq C\|u\| = C$. $\square$

Proposition 3.6 *Let f be a graphon and take any $\epsilon > 0$. Let $\mathscr{C}$ be a maximal collection of orthonormal eigenfunctions of K_f such that none of the eigenvalues $(\lambda_u)_{u \in \mathscr{C}}$ belong to $(-\epsilon, \epsilon)$. Then $|\mathscr{C}| \leq 1/\epsilon^2$ and for each $u \in \mathscr{C}$ and $x \in [0,1]$, $|u(x)| \leq 1/\epsilon$. Moreover, if*

$$g(x, y) := f(x, y) - \sum_{u \in \mathscr{C}} \lambda_u u(x) u(y),$$

then $\|K_g\| < \epsilon$.

Proof The bounds on $|\mathscr{C}|$ and $|u(x)|$ follow from Proposition 3.5. Since $g(x, y) = g(y, x)$, K_g is self-adjoint. Since f is bounded and the u's are bounded and $|\mathscr{C}|$ is finite, g is bounded. Thus, Proposition 3.4 implies that K_g is a compact operator. So by Proposition 2.10, there exist $v \in B_1([0,1]^2)$ and $\lambda \in \mathbb{R}$ such that $|\lambda| = \|K_g\|$ and $K_g v = \lambda v$.

Suppose that $|\lambda| \geq \epsilon$. Note that for any $u \in \mathscr{C}$, $K_g u = 0$, and therefore

$$(u, v) = \lambda^{-1}(u, K_g v) = \lambda^{-1}(K_g u, v) = 0.$$

Thus,

$$\lambda v = K_g v = K_f v - \sum_{u \in \mathscr{C}} \lambda_u (u, v) u = K_f v.$$

In other words, $\mathscr{C} \cup \{v\}$ is an orthonormal collection of eigenfunctions of K_f such that all eigenvalues are outside $(-\epsilon, \epsilon)$. This gives a contradiction, since $\mathscr{C}$ is a maximal collection with this property. $\square$

3.5 Compactness of the Space of Graphons

Given $n \geq 1$, let $\mathscr{W}_n$ be the set of all graphons that are constant in open squares of the form

$$\left(\frac{i-1}{n}, \frac{i}{n} \right) \times \left(\frac{j-1}{n}, \frac{j}{n} \right),$$

for every $1 \le i, j \le n$. Let $\mathcal{M}_n$ be the set of all measure-preserving bijections of $[0, 1]$ that linearly translate any interval of the form $((i - 1)/n, i/n)$ to another interval of the same type, and do not move points of the form i/n. Note that if $\sigma \in \mathcal{M}_n$ and $f \in \mathcal{W}_n$, then $f_\sigma \in \mathcal{W}_n$. Note also that $\mathcal{M}_n$ has a natural bijection with the symmetric group S_n and therefore $|\mathcal{M}_n| = n!$.

Let $\mathcal{U}_n$ be the set of all Borel measurable functions from $[0, 1]$ into $\mathbb{R}$ that are constant in intervals of the form $((i - 1)/n, i/n)$. Note that if $u \in \mathcal{U}_n$ and $\sigma \in \mathcal{M}_n$, then $u_\sigma \in \mathcal{U}_n$, where $u_\sigma(x) := u(\sigma x)$.

The goal of this section is to prove the following theorem, which says that the space $\widetilde{\mathcal{W}}$ defined in Sect. 3.1 is compact under the metric $\delta_\square$. The proof of this result will make use of the functional analytic results derived in Chap. 2 and the preceding section.

Theorem 3.1 (Weak Regularity Theorem for Graphons) *Given any $\epsilon \in (0, 1)$, there exists a set $\mathcal{W}(\epsilon) \subseteq \mathcal{W}$ with the following properties:*

(i) *There are universal constants C_1, C_2 and C_3 such that*

$$|\mathcal{W}(\epsilon)| \le C_1 \epsilon^{-C_2 \epsilon^{-C_3 \epsilon^{-2}}}.$$

(ii) *For any $f \in \mathcal{W}$, there exists $\sigma \in \mathcal{M}$ and $h \in \mathcal{W}(\epsilon)$ such that*

$$d_\square(f_\sigma, h) < \epsilon.$$

In particular, the metric space $(\widetilde{\mathcal{W}}, \delta_\square)$ defined in Sect. 3.1 is compact.

(iii) *If $f \in \mathcal{W}_n$, then the σ in part (ii) can be chosen to be in the set $\mathcal{M}_n$ defined above.*

Proof Throughout the proof, C, C_1, C_2 and C_3 will denote positive universal constants, whose values may change from line to line.

Fix $\epsilon > 0$. Take any $n \ge 1$ and a graphon $f \in \mathcal{W}_n$. Let $\mathcal{C}$ be a maximal collection of orthonormal eigenfunctions of K_f such that none of the eigenvalues belong to $(-\epsilon/5, \epsilon/5)$. Then by Proposition 3.6, $|\mathcal{C}| \le 25/\epsilon^2$ and for each $u \in \mathcal{C}$ and $x \in [0, 1]$, $|u(x)| \le 5/\epsilon$. Enumerate the elements of $\mathcal{C}$ as $\{u_1, \ldots, u_m\}$ and the corresponding eigenvalues as $\lambda_1, \ldots, \lambda_m$. Define

$$r(x, y) := \sum_{i=1}^{m} \lambda_i u_i(x) u_i(y).$$

Then again by Proposition 3.6,

$$\|K_f - K_r\| < \frac{\epsilon}{5}. \tag{3.5.1}$$

Let $A = A(\epsilon)$ be a discretization of the interval $[-5/\epsilon, 5/\epsilon]$ into a set of equally spaced points, so that the spacing between two neighboring points is $\leq \epsilon^5$. Clearly, one can construct A with $|A| \leq C\epsilon^{-6}$.

Similarly, let $B = B(\epsilon)$ be a discretization of $[-1, 1]$ into a grid of equally spaced points, so that the spacing between two neighboring points is $\leq \epsilon^6$. Then B can be chosen such that $|B| \leq C\epsilon^{-6}$.

For each i, let θ_i be the point in B that is closest to λ_i, breaking ties by some pre-fixed rule. For each i and x, let $v_i(x)$ be the point in A that is closest to $u_i(x)$. Let

$$s_1(x, y) := \sum_{i=1}^{m} \theta_i v_i(x) v_i(y).$$

Then for any x and y,

$$|r(x, y) - s_1(x, y)| \leq \sum_{i=1}^{m} |\lambda_i - \theta_i|(5/\epsilon)^2 + \sum_{i=1}^{m} |u_i(x) - v_i(x)|(5/\epsilon)$$

$$+ \sum_{i=1}^{m} |u_i(y) - v_i(y)|(5/\epsilon) \leq C\epsilon^2.$$

Consequently,

$$\|r - s_1\| \leq C\epsilon^2. \tag{3.5.2}$$

Since $f \in \mathcal{W}_n$, the eigenvalue equation $u_i = \lambda_i^{-1} K_f u_i$ implies that $u_i \in \mathcal{U}_n$ for every i. Consequently, v_i also belongs to $\mathcal{U}_n$ for each i. This allows a lexicographic reordering of the m-tuple of functions $(v_1, \ldots, v_m)$ by an element of $\mathcal{M}_n$, in the following sense. There exists $\sigma \in \mathcal{M}_n$ such that if $w_i(x) := v_i(\sigma x)$, then for any $0 \leq x \leq y \leq 1$ such that neither x nor y is of the form j/n for some integer j, the m-tuple $w(x) := (w_1(x), \ldots, w_m(x))$ comes before the m-tuple $w(y)$ in the lexicographic ordering. Define

$$s_2(x, y) = \sum_{i=1}^{m} \theta_i w_i(x) w_i(y) = s_1(\sigma x, \sigma y).$$

Then by (3.5.2), we get

$$\|r_\sigma - s_2\| \leq C\epsilon^2. \tag{3.5.3}$$

Let $\{a_1, a_2, \ldots, a_l\}$ be the lexicographic ordering of the elements of A^m, where $l = l(\epsilon, m) = |A|^m$. Then there are numbers $0 = n_0 \leq n_1 \leq n_2 \leq \cdots \leq n_l = n$ such that $w(x) = a_i$ when

$$\frac{n_{i-1}}{n} < x < \frac{n_i}{n}.$$

Note that if $n_{i-1} = n_i$ for some i, then it means that there is no x such that $w(x) = a_i$.

Let $k = k(\epsilon, m) := [l/\epsilon^{12}] + 1$, where $[x]$ denotes the integer part of x. For each $0 \leq i \leq l$, let k_i be the unique integer such that

$$\frac{k_i}{k} \leq \frac{n_i}{n} < \frac{k_i + 1}{k}.$$

Note that $0 = k_0 \leq k_1 \leq \cdots \leq k_l = k$. Define a new collection of functions $q_1, \ldots, q_m : [0, 1] \to \mathbb{R}$ as follows. For any $i \geq 1$ such that $k_{i-1} < k_i$, let

$$q(x) = (q_1(x), \ldots, q_m(x)) := a_i$$

for each $x \in (k_{i-1}/k, k_i/k)$. This covers all x except those of the form j/k. When $x = j/k$ for some integer j, define $q_i(x) = 0$ for each i.

Having defined $q_1, \ldots, q_m$, let

$$s_3(x, y) := \sum_{i=1}^{m} \theta_i q_i(x) q_i(y).$$

Take any $x \in [0, 1]$ that is not of the form j/k. Then there exists a unique i such that $x \in (k_{i-1}/k, k_i/k)$. If x also belongs to $(n_{i-1}/n, n_i/n)$, then $q(x) = w(x) = a_i$. On the other hand, if $x \notin (n_{i-1}/n, n_i/n)$, then x must belong to $(k_{i-1}/k, n_{i-1}/n]$, which is an interval of length $\leq 1/k$. Thus, the Lebesgue measure of the set of all x where $q(x) \neq w(x)$ is bounded above by l/k. Consequently, the Lebesgue measure of the set of all (x, y) where $s_3(x, y) \neq s_2(x, y)$ is bounded above by $2l/k$, since $s_3(x, y) \neq s_2(x, y)$ implies that $q(x) \neq w(x)$ or $q(y) \neq w(y)$.

Since $m \leq 25/\epsilon^2$, $|\theta_i| \leq 1$ for each i, and $|w_i(x)|$ and $|q_i(x)|$ are bounded by $5/\epsilon$ for each i and x, the absolute values of the functions s_2 and s_3 are both uniformly bounded above by $C\epsilon^{-4}$. Combining this with the observation made in the previous paragraph, we get

$$\|s_2 - s_3\| \leq C\epsilon^{-4}\sqrt{\frac{2l}{k}} \leq C\epsilon^2. \tag{3.5.4}$$

Note that the function s_3 is fully determined by the constants $\theta_1, \ldots, \theta_m$ and the functions $q_1, \ldots, q_m$. On the other hand, each θ_i is in B, and the functions $q_1, \ldots, q_m$ are completely determined by the numbers $k_1, \ldots, k_{l-1}$, which are integers between 0 and k. Combining these observations, we see that there is a finite set $\mathscr{W}'(\epsilon)$ of L^2 functions, depending only on ϵ, such that any s_3 obtained as above belongs to $\mathscr{W}'(\epsilon)$, and

$$|\mathscr{W}'(\epsilon)| \leq \sum_{m \leq 25/\epsilon^2} |B(\epsilon)|^m k(\epsilon, m)^{l(\epsilon, m)} \leq C_1 e^{-C_2 \epsilon^{-C_3 \epsilon^{-2}}}. \tag{3.5.5}$$

Combining (3.5.1), (3.5.3) and (3.5.4), and using the inequality (3.4.2), we get

$$\|K_{f_\sigma} - K_{s_3}\| \le \|K_{f_\sigma} - K_{r_\sigma}\| + \|K_{r_\sigma} - K_{s_2}\| + \|K_{s_2} - K_{s_3}\|$$
$$\le \|K_f - K_r\| + \|r_\sigma - s_2\| + \|s_2 - s_3\|$$
$$\le \frac{\epsilon}{5} + C\epsilon^2.$$

To summarize, what we have finished proving is the following. Given any $\epsilon \in (0, 1)$, we have constructed a set $\mathcal{W}'(\epsilon) \subseteq L^2([0, 1]^d)$ satisfying the bound (3.5.5) and the criterion that whenever $f \in \mathcal{W}_n$ for some n, there exists $\sigma \in \mathcal{M}_n$ and $g \in \mathcal{W}'(\epsilon)$ such that

$$\|K_{f_\sigma} - K_g\| \le \frac{\epsilon}{5} + C\epsilon^2,$$

where C is a positive universal constant.

Now take some arbitrary $f \in \mathcal{W}$ and $\epsilon \in (0, 1)$. Let $\hat{f}_n$ be the level n approximant of f, as defined in Sect. 2.2. Find n so large that $\|f - \hat{f}_n\| \le \epsilon/5$. Then by the conclusion summarized in the previous paragraph, there exist $g \in \mathcal{W}'(\epsilon)$ and $\sigma \in \mathcal{M}_n$ such that

$$\|K_{(\hat{f}_n)_\sigma} - K_g\| < \frac{\epsilon}{5} + C\epsilon^2.$$

Consequently, by (3.4.2),

$$\|K_{f_\sigma} - K_g\| \le \|K_{f_\sigma} - K_{(\hat{f}_n)_\sigma}\| + \|K_{(\hat{f}_n)_\sigma} - K_g\|$$
$$\le \|K_f - K_{\hat{f}_n}\| + \frac{\epsilon}{5} + C\epsilon^2$$
$$\le \|f - \hat{f}_n\| + \frac{\epsilon}{5} + C\epsilon^2 \le \frac{2\epsilon}{5} + C\epsilon^2.$$

Thus, there exists $\epsilon_0 \in (0, 1)$ such that the following is true when $0 < \epsilon \le \epsilon_0$. For any $f \in \mathcal{W}$, there exists $\sigma \in \mathcal{M}$ and $g \in \mathcal{W}'(\epsilon)$ such that

$$\|K_{f_\sigma} - K_g\| < \frac{\epsilon}{2}.$$

Moreover, if $f \in \mathcal{W}_n$, then σ can be chosen to be in $\mathcal{M}_n$.

For each $g \in \mathcal{W}'(\epsilon)$, choose an element $h \in \mathcal{W}$ such that $\|K_g - K_h\| < \epsilon/2$, provided that such an element exists. If there does not exist such an element, do not choose any h for this g. Let $\mathcal{W}(\epsilon)$ be the subset of $\mathcal{W}$ formed by pooling together the h's chosen in this manner. Then clearly $|\mathcal{W}(\epsilon)| \le |\mathcal{W}'(\epsilon)|$. Suppose that $\epsilon \le \epsilon_0$. Then we know that for any $f \in \mathcal{W}$, there exists $g \in \mathcal{W}'(\epsilon)$ and $\sigma \in \mathcal{M}$ such that

$\|K_{f_\sigma} - K_g\| < \epsilon/2$. In particular, this implies that there exists $h \in \mathscr{W}(\epsilon)$ such that $\|K_g - K_h\| < \epsilon/2$, and for this h,

$$\|K_{f_\sigma} - K_h\| \leq \|K_{f_\sigma} - K_g\| + \|K_g - K_h\| < \epsilon.$$

Moreover, we know that if $f \in \mathscr{W}_n$, then σ can be chosen to be in $\mathscr{M}_n$. By Lemma 3.1, this proves the theorem when $\epsilon \leq \epsilon_0$. If $\epsilon_0 < \epsilon < 1$, let $\mathscr{W}(\epsilon) = \mathscr{W}(\epsilon_0)$, and increase the value of C_1 in the statement of the theorem so that part (i) still remains valid. $\qquad\square$

Bibliographical Notes

Graph limit theory was formulated in a series of papers by Lovász and coauthors covering a span of several years. Some of the most important papers in this series are Lovász [14, 15], Lovász and Szegedy [18–21], Freedman et al. [10], Lovász and Sós [17] and Borgs et al. [4–6]. Austin [2] and Diaconis and Janson [9] trace some aspects of this theory back to the works of Aldous [1], Hoover [12] and Kallenberg [13] in the probability literature. Graph limit theory sheds light on various graph-theoretic topics such as graph homomorphisms, quasi-random graphs, property testing and extremal graph theory. A comprehensive survey is available in Lovász [16]. The theory has been developed for dense graphs (number of edges comparable to the square of number of vertices) but parallel theories for sparse graphs are beginning to emerge, for example in Bollobás and Riordan [3] and Borgs et al. [7, 8].

The definition of a graphon and the spaces $\mathscr{W}$ and $\widetilde{\mathscr{W}}$ were introduced by Borgs et al. [5]. The cut metric for graphs was introduced by Frieze and Kannan [11], and was extended to graphons in the above paper.

Theorem 3.1 is one of the key results of graph limit theory. Part (iii) of the theorem is essentially the weak regularity lemma of Frieze and Kannan [11]. This is a weak version of the famous regularity lemma of Szemerédi [23]. Part (ii) of the theorem is a restatement of Theorem 5.1 of Lovász and Szegedy [19], which converts the regularity lemma into a compactness theorem for $\widetilde{\mathscr{W}}$. I have not seen part (i) explicitly stated in the literature. The proof of Theorem 3.1 using the spectral properties of compact operators, as given here, is not the standard approach. The standard proof of Szemerédi's regularity lemma is combinatorial, and the passage to the compactness of $\widetilde{\mathscr{W}}$, as given in Lovász and Szegedy [19], uses martingales. The proof given here bears some similarities with the spectral techniques of Frieze and Kannan [11] and Szegedy [22]. I chose this route to give a self-contained presentation in a minimum number of pages.

References

1. Aldous, D. (1981). Representations for partially exchangeable arrays of random variables. *Journal of Multivariate Analysis, 11*(4), 581–598.
2. Austin, T. (2008). On exchangeable random variables and the statistics of large graphs and hypergraphs. *Probability Surveys, 5*, 80–145.
3. Bollobás, B., & Riordan, O. (2009). Metrics for sparse graphs. In *Surveys in combinatorics 2009*, vol. 365, London mathematical society of lecture note series (pp. 211–287). Cambridge: Cambridge University Press.
4. Borgs, C., Chayes, J., Lovász, L., Sós, V. T., & Vesztergombi, K. (2006). Counting graph homomorphisms. In *Topics in discrete mathematics*, vol. 26, Algorithms and combinatorics (pp. 315–371). Berlin: Springer.
5. Borgs, C., Chayes, J., Lovász, L., Sós, V. T., & Vesztergombi, K. (2008). Convergent sequences of dense graphs. I. Subgraph frequencies, metric properties and testing. *Advances in Mathematics, 219*(6), 1801–1851.
6. Borgs, C., Chayes, J., Lovász, L., Sós, V. T., & Vesztergombi, K. (2012). Convergent sequences of dense graphs. II. Multiway cuts and statistical physics. *Annals of Mathematics, 176*(1), 151–219.
7. Borgs, C., Chayes, J. T., Cohn, H., & Zhao, Y. (2014). An L^p theory of sparse graph convergence I: Limits, sparse random graph models, and power law distributions. *arXiv preprint arXiv:1401.2906*
8. Borgs, C., Chayes, J. T., Cohn, H., & Zhao, Y. (2014). An L^p theory of sparse graph convergence II: LD convergence, quotients, and right convergence. *arXiv preprint arXiv:1408.0744*
9. Diaconis, P., & Janson, S. (2008). Graph limits and exchangeable random graphs. *Rendiconti di Matematica e delle sue Applicazioni. Serie VII, 28*(1), 33–61. http://www1.mat.uniroma1.it/ricerca/rendiconti/
10. Freedman, M., Lovász, L., & Schrijver, A. (2007). Reflection positivity, rank connectivity, and homomorphism of graphs. *Journal of the American Mathematical Society, 20*, 37–51 (electronic).
11. Frieze, A., & Kannan, R. (1999). Quick approximation to matrices and applications. *Combinatorica, 19*, 175–220.
12. Hoover, D. N. (1982). Row-column exchangeability and a generalized model for probability. *Exchangeability in probability and statistics (Rome, 1981)*, (pp. 281–291). Amsterdam: North-Holland.
13. Kallenberg, O. (2005). *Probabilistic symmetries and invariance principles*. New York: Springer.
14. Lovász, L. (2006). The rank of connection matrices and the dimension of graph algebras. *European Journal of Combinatorics, 27*, 962–970.
15. Lovász, L. (2007). Connection matrices. In *Combinatorics, complexity, and chance*, vol. 34, Oxford lecture series in mathematics and its applications (pp. 179–190). Oxford: Oxford University Press.
16. Lovász, L. (2012). *Large networks and graph limits*. Providence: American Mathematical Society.
17. Lovász, L., & Sós, V. T. (2008). Generalized quasirandom graphs. *Journal of Combinatorial Theory, Series B, 98*, 146–163.
18. Lovász, L., & Szegedy, B. (2006). Limits of dense graph sequences. *Journal of Combinatorial Theory, Series B, 96*(6), 933–957.
19. Lovász, L., & Szegedy, B. (2007). Szemerédi's lemma for the analyst. *Geometric and Functional Analysis, 17*, 252–270.
20. Lovász, L., & Szegedy, B. (2010). Testing properties of graphs and functions. *Israel Journal of Mathematics, 178*, 113–156.
21. Lovász, L., & Szegedy, B. (2009). Contractors and connectors of graph algebras. *Journal of Graph Theory, 60* 11–30.

22. Szegedy, B. (2011). Limits of kernel operators and the spectral regularity lemma. *European Journal of Combinatorics, 32*(7), 1156–1167.
23. Szemerédi, E. (1978). Regular partitions of graphs. In *Problèmes combinatoires et théorie des graphes (Colloq. Internat. CNRS, Univ. Orsay, Orsay, 1976)*, pp. 399–401, Colloq. Internat. CNRS, 260, CNRS, Paris.

Chapter 4
Large Deviation Preliminaries

This chapter contains some basic definitions and results from abstract large deviation theory. As before, no background other than graduate-level probability and functional analysis is required, and the discussion is limited to the minimum required for the monograph.

4.1 Definition of Rate Function

Let $\mathscr{X}$ be a topological space and $\mathscr{B}$ be its Borel sigma-algebra. For any set $\Gamma \in \mathscr{B}$, let $\overline{\Gamma}$ denote the closure of Γ and let Γ° denote the interior of Γ. Let $\{\mu_n\}_{n\geq1}$ be a sequence of probability measures on $(\mathscr{X}, \mathscr{B})$. Let $I : \mathscr{X} \to [-\infty, \infty]$ be a Borel measurable function and $\{\epsilon_n\}_{n\geq1}$ be a sequence of positive real numbers tending to zero.

Definition 4.1 The sequence of probability measures $\{\mu_n\}_{n\geq1}$ is said to satisfy a large deviation principle (LDP) with rate ϵ_n and rate function I if for all $\Gamma \in \mathscr{B}$,

$$- \inf_{x\in\Gamma^{\circ}} I(x) \leq \liminf_{n\to\infty} \epsilon_n \log \mu_n(\Gamma) \leq \limsup_{n\to\infty} \epsilon_n \log \mu_n(\Gamma) \leq - \inf_{x\in\overline{\Gamma}} I(x).$$

The goal of large deviation theory is to understand probabilities of rare events by calculating rate functions for sequences of probability measures. If the reader is unfamiliar with the basic ideas of large deviation theory, the following simple exercise may help clarify the meaning of a rate function.

Exercise 4.1 Let μ_n be the Gaussian distribution with mean zero and variance $1/n$ on $\mathbb{R}$. Show that μ_n satisfies a large deviation principle with rate $1/n$ and rate function $x^2/2$.

© Springer International Publishing AG 2017
S. Chatterjee, *Large Deviations for Random Graphs*, Lecture Notes
in Mathematics 2197, DOI 10.1007/978-3-319-65816-2_4

4.2 A Local-to-Global Transference Principle

When trying to prove a large deviation upper bound for a compact set Γ, it suffices to prove it in arbitrarily small balls. This is the content of the following useful lemma.

Lemma 4.1 *Let $\{\mu_n\}_{n\geq 1}$ be a sequence of probability measures on a metric space $\mathscr{X}$ (equipped with its Borel sigma-algebra), and $\{\epsilon_n\}_{n\geq 1}$ be a sequence of positive real numbers tending to zero. Let $B(x, \eta)$ denote the closed ball of radius around a point x, and let $\Gamma \subseteq \mathscr{X}$ be a compact set. Suppose that $I : \Gamma \to [-\infty, \infty]$ is a function such that for every $x \in \Gamma$,*

$$\lim_{\eta \to 0} \limsup_{n \to \infty} \epsilon_n \log \mu_n(B(x, \eta)) \leq -I(x).$$

Then for any closed set $F \subseteq \Gamma$,

$$\limsup_{n \to \infty} \epsilon_n \log \mu_n(F) \leq -\inf_{x \in F} I(x).$$

Proof Take any closed set $F \subseteq \Gamma$. If F is empty, then there is nothing to prove. So assume that F is nonempty, and assume that the first inequality holds. Fix $\eta > 0$ and use the compactness of Γ to pick $x_1, \ldots, x_k \in F$ such that

$$F \subseteq \bigcup_{i=1}^{k} B(x_i, \eta).$$

Then the union bound for probability measures implies that

$$\limsup_{n \to \infty} \epsilon_n \log \mu_n(F) \leq \max_{1 \leq i \leq k} \limsup_{n \to \infty} \epsilon_n \log \mu_n(B(x_i, \eta))$$

$$\leq \sup_{x \in F} \limsup_{n \to \infty} \epsilon_n \log \mu_n(B(x, \eta)).$$

Therefore it suffices to prove that

$$\limsup_{\eta \to 0} \sup_{x \in F} \limsup_{n \to \infty} \epsilon_n \log \mu_n(B(x, \eta)) \leq \sup_{x \in F}(-I(x)).$$

Suppose that this is false. Then there exists a sequence $\{x_k\}_{k \geq 1}$ in F and a sequence $\{\eta_k\}_{k \geq 1}$ of positive numbers tending to zero such that

$$\lim_{k \to \infty} \limsup_{n \to \infty} \epsilon_n \log \mu_n(B(x_k, \eta_k)) > \sup_{x \in F}(-I(x)). \tag{4.2.1}$$

By the compactness of F, we may assume, after passing to a subsequence if necessary, that $x_k \to x \in F$. Let

$$\eta_k' := \eta_k + d(x_k, x),$$

where d is the metric on $\mathscr{X}$. Then $\eta_k' \to 0$ and $B(x_k, \eta_k) \subseteq B(x, \eta_k')$. Therefore, by the assumed inequality in the statement of the lemma and the inequality (4.2.1),

$$
\begin{aligned}
-I(x) &\geq \lim_{k\to\infty} \limsup_{n\to\infty} \epsilon_n \log \mu_n(B(x, \eta_k')) \\
&\geq \lim_{k\to\infty} \limsup_{n\to\infty} \epsilon_n \log \mu_n(B(x_k, \eta_k)) \\
&> \sup_{x\in F}(-I(x)),
\end{aligned}
$$

which is a contradiction. $\qquad\qquad\qquad\qquad\qquad\qquad\qquad\qquad\qquad\qquad\qquad\quad$ $\square$

4.3 A General Upper Bound

Recall that a topological vector space is a vector space endowed with a topology under which the vector space operations are continuous. Let $\mathscr{X}$ be a real topological vector space whose topology has the Hausdorff property. Let $\mathscr{X}^*$ be the dual space of $\mathscr{X}$, that is, the space of all continuous linear functionals on $\mathscr{X}$. Let $\mathscr{B}$ be the Borel sigma-algebra of $\mathscr{X}$ and let $\{\mu_n\}_{n\geq 1}$ be a sequence of probability measures on $(\mathscr{X}, \mathscr{B})$. Define the logarithmic moment generating function $\Lambda_n : \mathscr{X}^* \to (-\infty, \infty]$ of μ_n as

$$\Lambda_n(\lambda) := \log \int_{\mathscr{X}} e^{\lambda(x)} d\mu_n(x).$$

Let $\{\epsilon_n\}_{n\geq 1}$ be a sequence of positive real numbers tending to zero. Define a function $\bar{\Lambda} : \mathscr{X}^* \to [-\infty, \infty]$ as

$$\bar{\Lambda}(\lambda) := \limsup_{n\to\infty} \epsilon_n \Lambda_n(\lambda/\epsilon_n). \qquad\qquad (4.3.1)$$

The Fenchel–Legendre transform of $\bar{\Lambda}$ is the function $\bar{\Lambda}^* : \mathscr{X} \to [-\infty, \infty]$ defined as

$$\bar{\Lambda}^*(x) := \sup_{\lambda\in\mathscr{X}^*} (\lambda(x) - \bar{\Lambda}(\lambda)). \qquad\qquad (4.3.2)$$

The following result shows that $\bar{\Lambda}^*$ is an upper bound for the rate function of μ_n if the μ_n's are supported on a compact set. This is one of the commonly used tools in large deviation theory.

Theorem 4.1 *For any compact set* $\Gamma \subseteq \mathscr{X}$,

$$\limsup_{n \to \infty} \epsilon_n \log \mu_n(\Gamma) \leq - \inf_{x \in \Gamma} \bar{\Lambda}^*(x).$$

Proof Fix a compact set $\Gamma \subseteq \mathscr{X}$ and a number $\delta > 0$. Let

$$I^\delta(x) := \min\{\bar{\Lambda}^*(x) - \delta, \ 1/\delta\}.$$

The definition (4.3.2) of $\bar{\Lambda}^*$ shows that for any $x \in \Gamma$, there exists $\lambda_x \in \mathscr{X}^*$ such that

$$I^\delta(x) \leq \lambda_x(x) - \bar{\Lambda}(\lambda_x). \tag{4.3.3}$$

Since λ_x is a continuous linear functional, there exists an open neighborhood A_x of x such that

$$\inf_{y \in A_x} (\lambda_x(y) - \lambda_x(x)) \geq -\delta. \tag{4.3.4}$$

For any $\theta \in \mathscr{X}^*$,

$$\mu_n(A_x) \leq \int_{\mathscr{X}} \frac{e^{\theta(y) - \theta(x)}}{e^{\inf_{z \in A_x}(\theta(z) - \theta(x))}} d\mu_n(y).$$

Taking $\theta = \lambda_x/\epsilon_n$ and using (4.3.4) gives

$$\epsilon_n \log \mu_n(A_x) \leq \delta - \lambda_x(x) + \epsilon_n \Lambda_n(\lambda_x/\epsilon_n). \tag{4.3.5}$$

Since Γ is compact, there exists a finite collection of points $x_1, \ldots, x_N \in \Gamma$ such that

$$\Gamma \subseteq \bigcup_{i=1}^{N} A_{x_i},$$

and hence

$$\mu_n(\Gamma) \leq \sum_{i=1}^{N} \mu_n(A_{x_i}) \leq N \max_{1 \leq i \leq N} \mu_n(A_{x_i}).$$

Therefore by (4.3.5),

$$\epsilon_n \log \mu_n(\Gamma) \le \epsilon_n \log N + \delta - \min_{1 \le i \le N} (\lambda_{x_i}(x_i) - \epsilon_n \Lambda_n(\lambda_{x_i}/\epsilon_n)).$$

By the definition (4.3.1) of $\bar{\Lambda}$ and the inequality (4.3.3), this gives

$$\limsup_{n \to \infty} \epsilon_n \log \mu_n(\Gamma) \le \delta - \min_{1 \le i \le N} (\lambda_{x_i}(x_i) - \bar{\Lambda}(x_i))$$

$$\le \delta - \min_{1 \le i \le N} I^\delta(x_i)$$

$$\le \delta - \inf_{x \in \Gamma} I^\delta(x).$$

The proof is completed by taking $\delta \to 0$. $\qquad\square$

For the uninitiated reader, the following exercises may help clarify the nature of the upper bound from Theorem 4.1.

Exercise 4.2 Let $X_1, X_2, \ldots$ be i.i.d. random variables that take value 1 with probability $1/2$ and -1 with probability $1/2$. Let μ_n be the probability law of the sample mean $(X_1 + \cdots + X_n)/n$. With $\mathscr{X} = \mathbb{R}$ and $\epsilon_n = 1/n$, compute $\bar{\Lambda}^*$ in the problem.

Exercise 4.3 Let μ_n be as in Exercise 4.2. Show that the function $\bar{\Lambda}^*$ for this problem is in fact the large deviation rate function for the sequence $\{\mu_n\}_{n \ge 1}$.

4.4 The Azuma–Hoeffding Inequality

In this section we will state and prove a useful probability inequality, known as the Azuma–Hoeffding inequality. In contrast with Theorem 4.1, it gives a finite sample bound with no limits.

Recall that a filtration of sigma-algebras $\{\mathscr{F}_i\}_{i \ge 0}$ is a sequence of sigma-algebras such that $\mathscr{F}_i \subseteq \mathscr{F}_{i+1}$ for each i. Given a filtration $\{\mathscr{F}_i\}_{i \ge 0}$ on a probability space, with $\mathscr{F}_0$ being the trivial sigma-algebra, recall that a sequence of random variables $\{X_i\}_{i \ge 1}$ defined on this space is called a martingale difference sequence if X_i is $\mathscr{F}_i$-measurable, $\mathbb{E}|X_i| < \infty$ and $\mathbb{E}(X_i \mid \mathscr{F}_{i-1}) = 0$ for each $i \ge 1$. (I am assuming that the reader is familiar with the abstract definition of conditional expectation.)

Theorem 4.2 (Azuma–Hoeffding Inequality) *Suppose that $\{X_i\}_{1 \le i \le n}$ is a martingale difference sequence with respect to some filtration $\{\mathscr{F}_i\}_{0 \le i \le n}$ with $\mathscr{F}_0 =$ the trivial sigma-algebra. Suppose that $A_1, \ldots, A_n, B_1, \ldots, B_n$ are random variables and $c_1, \ldots, c_n$ are constants such that for each i, A_i and B_i are $\mathscr{F}_{i-1}$-measurable, and with probability one, $A_i \le X_i \le B_i$ and $B_i - A_i \le c_i$. Let $S_n := X_1 + \cdots + X_n$.*

Then for any x ≥ 0,

$$\max\{\mathbb{P}(S_n \geq x), \mathbb{P}(S_n \leq -x)\} \leq \exp\left(-\frac{2x^2}{\sum_{i=1}^n c_i^2}\right).$$

The key ingredient in the proof of the Azuma–Hoeffding inequality is Hoeffding's lemma, stated below.

Lemma 4.2 (Hoeffding's Lemma) *Let X be a random variable that is bounded between two constants a and b with probability one. If $\mathbb{E}(X) = 0$, then for any $\theta \geq 0$,*

$$\mathbb{E}(e^{\theta X}) \leq e^{\theta^2 (b-a)^2 / 8}.$$

Proof Without loss of generality, suppose that $a < 0 < b$. For any $a < x < b$ and $\theta \in \mathbb{R}$,

$$e^{\theta x} = e^{t\theta b + (1-t)\theta a},$$

where

$$t = \frac{x-a}{b-a} \in [0, 1].$$

Therefore by Jensen's inequality (Proposition 2.1),

$$e^{\theta x} \leq t e^{\theta b} + (1-t) e^{\theta a}.$$

Since $\mathbb{E}(X) = 0$, this implies that

$$\mathbb{E}(e^{\theta X}) \leq \frac{b e^{\theta a} - a e^{\theta b}}{b-a}. \tag{4.4.1}$$

For $0 \leq t \leq 1$, let

$$h(t) := \log\left(\frac{b e^{t\theta a} - a e^{t\theta b}}{b-a}\right)$$

$$= t\theta a + \log(1 - p + p e^{t\theta(b-a)}),$$

where

$$p := \frac{-a}{b-a} \in (0, 1).$$

An easy verification shows that $h(0) = h'(0) = 0$. Furthermore,

$$h''(t) = \theta^2(b-a)^2 y(1-y),$$

where

$$y = \frac{e^{t\theta(b-a)}}{1 - p + pe^{t\theta(b-a)}} \in (0, 1).$$

This shows that $h''(t) \leq \theta^2(b-a)^2/4$ for every $t \in [0, 1]$. Together with the inequality (4.4.1), this completes the proof of the lemma.   □

Armed with Hoeffding's lemma, we can now easily prove the Azuma–Hoeffding inequality.

Proof (Proof of Theorem 4.2) Let $m_i(\theta) := \mathbb{E}(e^{\theta X_i} \mid \mathscr{F}_{i-1})$. Then by the given conditions and Lemma 4.2,

$$m_i(\theta) \leq e^{\theta^2 c_i^2/8}.$$

Let $\phi_i(\theta) := \mathbb{E}(e^{\theta S_i})$, where $S_i = X_1 + \cdots + X_i$. Then the above inequality implies that

$$\phi_n(\theta) = \mathbb{E}(e^{\theta S_{n-1}} m_n(\theta)) \leq \phi_{n-1}(\theta) e^{\theta^2 c_n^2/8}.$$

Proceeding inductively, we get

$$\phi_n(\theta) \leq e^{\theta^2(c_1^2 + \cdots + c_n^2)/8}.$$

Therefore Chebychev's inequality implies that for any $x \geq 0$ and $\theta \geq 0$,

$$\mathbb{P}(S_n \geq x) \leq \mathbb{P}(e^{\theta S_n} \geq e^{-\theta x})$$

$$\leq e^{-\theta x} \phi_n(\theta) \leq e^{-\theta x + \theta^2(c_1^2 + \cdots + c_n^2)/8}.$$

Choosing $\theta = 4x/(c_1^2 + \cdots + c_n^2)$ gives

$$\mathbb{P}(S_n \geq x) \leq \exp\left(-\frac{2x^2}{c_1^2 + \cdots + c_n^2}\right).$$

By a symmetrical argument, the same bound is proved for $\mathbb{P}(S_n \leq -x)$. The proof is completed by combining the two bounds.   □

4.5   McDiarmid's Inequality

The Azuma–Hoeffding inequality has the following important corollary, which is known as McDiarmid's inequality or the bounded differences inequality.

Theorem 4.3 (McDiarmid's Inequality)   *Let $\mathscr{X}$ be a set endowed with a sigma-algebra. Let $X_1, \ldots, X_n$ be independent $\mathscr{X}$-valued random variables. Let $X_1', \ldots, X_n'$ be independent copies of $X_1, \ldots, X_n$. Suppose that $f : \mathscr{X}^n \to \mathbb{R}$ is a measurable function and $c_1, \ldots, c_n$ are constants such that for each i, with probability one,*

$$|f(X_1, \ldots, X_n) - f(X_1, \ldots, X_i', \ldots, X_n)| \le c_i.$$

Let $W := f(X_1, \ldots, X_n)$. Then for any $x \ge 0$,

$$\max\{\mathbb{P}(W - \mathbb{E}(W) \ge x),\, \mathbb{P}(W - \mathbb{E}(W) \le -x)\} \le \exp\left(-\frac{2x^2}{\sum_{i=1}^n c_i^2}\right).$$

Proof Let $\mathscr{F}_0$ be the trivial sigma-algebra, and let $\mathscr{F}_i$ be the sigma-algebra generated by $X_1, \ldots, X_i$. Let

$$Y_i := \mathbb{E}(W \mid \mathscr{F}_i) - \mathbb{E}(W \mid \mathscr{F}_{i-1}),$$

so that $Y_1, \ldots, Y_n$ is a martingale difference sequence and $Y_1 + \cdots + Y_n = W - \mathbb{E}(W)$. Thus, we are in the setting of the Azuma–Hoeffding inequality.

Take any i. Let $W_i := f(X_1, \ldots, X_i', \ldots, X_n)$ and let $\mathscr{G}_i$ be the sigma-algebra generated by $(X_1, \ldots, X_i', \ldots X_n)$. Let U_i and L_i denote the essential supremum and essential infimum of $W - W_i$ given $\mathscr{G}_i$. By the given condition on f, conditional on $\mathscr{G}_i$, the maximum possible variation of W as X_i varies is bounded above by c_i. Thus, $U_i - L_i \le c_i$.

Now, since $L_i \le W - W_i \le U_i$ with probability one, it follows that

$$A_i \le \mathbb{E}(W - W_i \mid \mathscr{F}_i) \le B_i$$

with probability one, where $A_i = \mathbb{E}(L_i \mid \mathscr{F}_i)$ and $B_i = \mathbb{E}(U_i \mid \mathscr{F}_i)$. Clearly, $B_i - A_i \le c_i$. By the independence of the X_i's, A_i and B_i are $\mathscr{F}_{i-1}$-measurable. Also by independence, $\mathbb{E}(W_i \mid \mathscr{F}_i) = \mathbb{E}(W \mid \mathscr{F}_{i-1})$, and therefore $\mathbb{E}(W - W_i \mid \mathscr{F}_i) = Y_i$. The proof is now completed by an application of the Azuma–Hoeffding inequality (Theorem 4.2).   □

Bibliographical Notes

The materials contained in Sects. 4.1 and 4.3 are extracted from the textbook of Dembo and Zeitouni [3], with minor notational modifications. Adhering to the general principle being followed in this monograph, I have presented only as much as necessary. For example, I have avoided discussing general lower bounds, because the large deviation lower bounds required in this monograph will be derived directly without appealing to the general theory.

The abstract framework for large deviations in topological spaces was formulated by Varadhan [9]. The idea behind the general upper bound of Theorem 4.1 goes back to results of Cramér and Chernoff for sums of i.i.d. random variables. An early version of the general result under additional assumptions was given in Gärtner [4]. The version presented here was proved by Stroock [8] and de Acosta [2].

Hoeffding's lemma and the Azuma–Hoeffding inequality for sums of independent random variables was proved by Hoeffding [5]. It was later generalized by Azuma [1] to sums of martingale differences. It took a while for mathematicians to realize the usefulness of the Azuma–Hoeffding inequality, until it was pointed out by McDiarmid [6]. The central idea behind McDiarmid's inequality was, however, already exploited in the earlier work of Shamir and Spencer [7].

References

1. Azuma, K. (1967). Weighted sums of certain dependent random variables. *Tohoku Mathematical Journal, Second Series, 19*(3), 357–367.
2. de Acosta, A. (1985). Upper bounds for large deviations of dependent random vectors. *Zeitschrift für Wahrscheinlichkeitstheorie und Verwandte Gebiete, 69*, 551–565.
3. Dembo, A., & Zeitouni, O. (2010). *Large deviations techniques and applications.* Corrected reprint of the second (1998) edition. Berlin: Springer.
4. Gärtner, J. (1977). On large deviations from the invariant measure. *Theory of Probablity and its Applications, 22*, 24–39.
5. Hoeffding, W. (1963). Probability inequalities for sums of bounded random variables. *Journal of the American Statistical Association, 58*, 13–30.
6. McDiarmid, C. (1989). On the method of bounded differences. In J. Siemons (Ed.), *Surveys in combinatorics.* London Mathematical Society Lecture Notes Series, vol. 141, pp. 148–188. Cambridge: Cambridge University Press.
7. Shamir, E., & Spencer, J. (1987). Sharp concentration of the chromatic number on random graphs $G_{n,p}$. *Combinatorica, 7*(1), 121–129.
8. Stroock, D. W. (1984). *An introduction to the theory of large deviations.* Berlin: Springer.
9. Varadhan, S. R. S. (1966). Asymptotic probabilities and differential equations. *Communications on Pure and Applied Mathematics, 19*, 261–286.

Chapter 5
Large Deviations for Dense Random Graphs

A dense graph is a graph whose number of edges is comparable to the square of the number of vertices. The main result of this chapter is the formulation and proof of the large deviation principle for dense Erdős–Rényi random graphs. We will see later that this result can be used to derive large deviation principles for a large class of models. These and other applications will be given in later chapters. The results and definitions from the previous chapters will be used extensively in the proofs of this chapter.

5.1 The Erdős–Rényi Model

Let n be a positive integer and p be an element of $[0, 1]$. The Erdős–Rényi model defines an undirected random graph on the vertex set $\{1, 2, \ldots, n\}$ by declaring that any two vertices are connected by an edge with probability p, and that these assignments are independent of each other. The resulting random graph model is denoted by $G(n, p)$.

Let G be an Erdős–Rényi random graph with parameters n and p, defined on an abstract probability space $(\Omega, \mathcal{F}, \mathbb{P})$. Recall the definition of the graphon f^G of G from Chap. 3. Recall the space $\mathscr{W}$ and the cut metric on this space. Equip $\mathscr{W}$ with the Borel sigma-algebra induced by the cut metric, and define a probability $\mathbb{P}_{n,p}$ on this space as

$$\mathbb{P}_{n,p}(B) := \mathbb{P}(f^G \in B) \tag{5.1.1}$$

for every Borel measurable subset B of $\mathscr{W}$. In other words, $\mathbb{P}_{n,p}$ is the probability measure on $\mathscr{W}$ induced by the random graph G. Since f^G can take only finitely many values, the event $\{f^G \in B\}$ is guaranteed to be measurable.

© Springer International Publishing AG 2017
S. Chatterjee, *Large Deviations for Random Graphs*, Lecture Notes in Mathematics 2197, DOI 10.1007/978-3-319-65816-2_5

Let $\widetilde{f^G}$ be the image of f^G in the quotient space $\widetilde{\mathscr{W}}$. When $\widetilde{\mathscr{W}}$ is equipped with the Borel sigma-algebra induced by the metric $\delta_\square$, the random element $\widetilde{f^G}$ induces a probability measure $\widetilde{\mathbb{P}}_{n,p}$ the same way that $\mathbb{P}_{n,p}$ was defined:

$$\widetilde{\mathbb{P}}_{n,p}(\widetilde{B}) = \mathbb{P}(\widetilde{f^G} \in \widetilde{B}) \tag{5.1.2}$$

for every Borel set $\widetilde{B} \subseteq \widetilde{\mathscr{W}}$. Again, since $\widetilde{f^G}$ can take only finitely many values, there are no measurability issues. The following exercise describes the asymptotic behavior of $\widetilde{\mathbb{P}}_{n,p}$.

Exercise 5.1 Prove that if $\tilde{p}$ denotes the element of $\widetilde{\mathscr{W}}$ representing the graphon that is identically equal to p, and G is an Erdős–Rényi $G(n, p)$ graph, then $\delta_\square(\widetilde{f^G}, \tilde{p}) \to 0$ in probability as $n \to \infty$. Equivalently, show that for any open set $\widetilde{U}$ containing $\tilde{p}$, $\widetilde{\mathbb{P}}_{n,p}(\widetilde{U}) \to 1$ as $n \to \infty$.

5.2   The Rate Function

Given $p \in (0, 1)$, let $I_p : [0, 1] \to \mathbb{R}$ be the function

$$I_p(u) := u \log \frac{u}{p} + (1 - u) \log \frac{1 - u}{1 - p}. \tag{5.2.1}$$

Lemma 5.1 *The function I_p can be alternately expressed as*

$$I_p(u) = \sup_{a\in\mathbb{R}}(au - \log(pe^a + 1 - p)).$$

Proof Fixing u, let the term within the supremum be denoted by $J(a)$. When $u \in (0, 1)$, it is an easy calculus exercise to verify that J is concave and $J(a) \to \infty$ as $a \to \pm\infty$, and the maximum of J is indeed $I_p(u)$. When $u = 1$, $J(a) < I_p(1)$ for all $a < \infty$ and $J(a) \to I_p(1)$ as $a \to \infty$. Similarly, when $u = 0$, $J(a) < I_p(0)$ for all $a > -\infty$ and $J(a) \to I_p(0)$ as $a \to -\infty$. $\qquad\square$

The domain of the function I_p can be extended to $\mathscr{W}$ by defining

$$I_p(h) := \int_{[0,1]^2} I_p(h(x, y))\, dx\, dy.$$

We would now like to show that I_p is a lower semi-continuous function on $\mathscr{W}$. Recall that a function f from a topological space $\mathscr{X}$ into $\mathbb{R}$ is called lower semi-continuous if $f^{-1}((a, \infty))$ is open for every $a \in \mathbb{R}$. Clearly, continuous functions are lower semi-continuous. From the definition it follows easily that if $\{f_\alpha\}_{\alpha\in\mathscr{A}}$ is an arbitrary collection of lower semi-continuous functions on $\mathscr{X}$, then $f := \sup_{\alpha\in\mathscr{A}} f_\alpha$ is again lower semi-continuous, because

$$f^{-1}((a, \infty)) = \bigcup_{\alpha \in \mathscr{A}} f_\alpha^{-1}((a, \infty)).$$

An important characterization of lower semi-continuous functions on metric spaces says that if $\mathscr{X}$ is a metric space, then a function $f : \mathscr{X} \to \mathbb{R}$ is lower semi-continuous if and only if for every sequence $\{x_n\}_{n \geq 1}$ in $\mathscr{X}$ that converges to a point x,

$$\liminf_{n \to \infty} f(x_n) \geq f(x).$$

An easy consequence of the above inequality is that on compact metric spaces, a lower semi-continuous function f must necessarily attain its minimum, and moreover the set of minima is compact.

Lemma 5.2 *Let S be the set of $a \in L^2([0, 1]^2)$ that satisfy the symmetry condition $a(x, y) = a(y, x)$. The function I_p on $\mathscr{W}$ can be alternately expressed as*

$$I_p(h) = \sup_{a \in S} J_{p,a}(h),$$

where

$$J_{p,a}(h) := \int_{[0,1]^2} (a(x, y)h(x, y) - \log(pe^{a(x,y)} + 1 - p)) \, dx \, dy.$$

Proof Lemma 5.1 implies that

$$I_p(h) \geq \sup_{a \in S} J_{p,a}(h).$$

For the opposite inequality, let

$$a^*(x, y) := \log \frac{h(x, y)}{p} - \log \frac{1 - h(x, y)}{1 - p},$$

and notice that

$$I_p(h(x, y)) = a^*(x, y)h(x, y) - \log(pe^{a^*(x,y)} + 1 - p). \tag{5.2.2}$$

This would suffice to complete the proof if a^* were guaranteed to be in L^2. However, there is no such guarantee, and therefore we need to work a bit more. For each $\epsilon \in (0, 1)$, let

$$A_\epsilon := \{(x, y) \in [0, 1]^2 : \epsilon \leq h(x, y) \leq 1 - \epsilon\},$$

$$B_\epsilon := \{(x, y) \in [0, 1]^2 : 0 < h(x, y) < \epsilon \text{ or } 1 - \epsilon < h(x, y) < 1\},$$

and define

$$E := \{(x, y) \in [0, 1]^2 : h(x, y) = 1\},$$
$$F := \{(x, y) \in [0, 1]^2 : h(x, y) = 0\}.$$

For each $\epsilon \in (0, 1)$ and $M \geq 1$, define

$$a_{M,\epsilon}(x, y) := \begin{cases} a^*(x, y) & \text{if } (x, y) \in A_\epsilon, \\ 0 & \text{if } (x, y) \in B_\epsilon, \\ M & \text{if } (x, y) \in E, \\ -M & \text{if } (x, y) \in F. \end{cases}$$

Clearly, $a_{M,\epsilon}$ is bounded, and therefore in L^2. Also, it is symmetric, since h is symmetric. Let

$$K_{M,\epsilon}(x, y) := I_p(h(x, y)) - (a_{M,\epsilon}(x, y)h(x, y) - \log(pe^{a_{M,\epsilon}(x,y)} + 1 - p)).$$

Then by (5.2.2), $K_{M,\epsilon} = 0$ on A_ϵ. It is easy to verify that I_p is a bounded function on $[0, 1]$. Therefore $|K_{M,\epsilon}| \leq C(p)$ on B_ϵ, where $C(p)$ is a constant that depends only on p. Next, note that if $(x, y) \in E$, then

$$K_{M,\epsilon}(x, y) = \log \frac{1}{p} - (M - \log(pe^M + 1 - p))$$

$$= \log \frac{1}{p} - \log \frac{1}{p + (1 - p)e^{-M}}.$$

Similarly, if $(x, y) \in F$, then

$$K_{M,\epsilon}(x, y) = \log \frac{1}{1 - p} - \log \frac{1}{pe^{-M} + 1 - p}.$$

Therefore $|K_{M,\epsilon}| \leq C(p, M)$ on the set $E \cup F$, where $C(p, M)$ is a constant that tends to zero as $M \to \infty$. Combining the above observations, we get

$$\int_{[0,1]^2} |K_{M,\epsilon}(x, y)| \, dx \, dy \leq C(p)\mathrm{Leb}(B_\epsilon) + C(p, M),$$

where $\mathrm{Leb}(B_\epsilon)$ is the Lebesgue measure of B_ϵ. Consequently,

$$I_p(h) \leq \sup_{a \in L^2([0,1]^2)} J_{p,a}(h) + C(p)\mathrm{Leb}(B_\epsilon) + C(p, M).$$

Note that B_ϵ are bounded sets that decrease monotonically as $\epsilon \downarrow 0$, and the intersection of these sets is empty. Therefore $\mathrm{Leb}(B_\epsilon) \to 0$ as $\epsilon \to 0$. Also, $C(p, M) \to 0$ as $M \to \infty$. This completes the proof. $\qquad\square$

The main point of the representation given by the above lemma is to prove the following corollary.

Corollary 5.1 *The function I_p is lower semi-continuous with respect to the topology induced by the cut metric $d_\square$ on $\mathscr{W}$.*

Proof By the definition of the weak topology (Chap. 2, Sect. 2.3), the maps $J_{p,a}$ are continuous under the weak topology for each square-integrable a. Since the cut topology is stronger than the weak topology (Chap. 3, Proposition 3.3), these maps are continuous under the cut topology as well. Therefore by Lemma 5.2, I_p is lower semi-continuous under the cut topology on $\mathscr{W}$. $\qquad\square$

A consequence of the above corollary is the following proposition, that allows us to define the rate function on $\widetilde{\mathscr{W}}$. Recall the pseudometric $\delta_\square$ and the set $\mathscr{M}$ of measure preserving bijections defined in Chap. 3, Sect. 3.1.

Proposition 5.1 *If $\delta_\square(f, g) = 0$ for some $f, g \in \mathscr{W}$, then $I_p(f) = I_p(g)$. Consequently, it is valid to define I_p on $\widetilde{\mathscr{W}}$ as $I_p(\tilde{h}) := I_p(h)$. Moreover, with this definition, I_p is lower semi-continuous on $\widetilde{\mathscr{W}}$.*

Proof If $\delta_\square(f, g) = 0$, then there is a sequence $\{\sigma_n\}_{n \geq 1}$ in $\mathscr{M}$ such that

$$\lim_{n\to\infty} d_\square(f, g_{\sigma_n}) \to 0.$$

Therefore by Corollary 5.1 and Exercise 3.3,

$$I_p(f) \leq \liminf_{n\to\infty} I_p(g_{\sigma_n}) = I_p(g).$$

By symmetry, $I_p(g) \leq I_p(f)$. Lower semi-continuity of I_p on $\widetilde{\mathscr{W}}$ follows easily from lower semi-continuity on $\mathscr{W}$. $\qquad\square$

5.3 Large Deviation Upper Bound in the Weak Topology

Fix $p \in (0, 1)$. The goal of this section is to prove a large deviation upper bound in the weak topology for the sequence of measures $\{\mathbb{P}_{n,p}\}_{n \geq 1}$ defined in Sect. 5.1. This is not the main objective of this chapter, but a stepping stone to the more powerful result that will be proved in the next section.

Theorem 5.1 *Let $\mathbb{P}_{n,p}$ be defined as in Sect. 5.1 and I_p be defined as in Sect. 5.2. Then for every weakly closed set $F \subseteq \mathscr{W}$,*

$$\limsup_{n\to\infty} \frac{2}{n^2} \log \mathbb{P}_{n,p}(F) \leq -\inf_{f\in F} I_p(f).$$

Proof Let $\mathscr{X}$ be the vector space $L^2([0, 1]^2)$ with the weak topology. For each $a \in \mathscr{X}$ and $f \in \mathscr{X}$, let

$$\lambda_a(f) := (a, f) = \int_{[0,1]^2} a(x, y) f(x, y) \, dx \, dy.$$

Then $\lambda_a \in \mathscr{X}^*$. Define $\Lambda_n : \mathscr{X}^* \to \mathbb{R}$ as

$$\Lambda_n(\lambda) := \log \int_{\mathscr{X}} e^{\lambda(f)} d\mathbb{P}_{n,p}(f),$$

and let

$$\bar{\Lambda}(\lambda) := \limsup_{n \to \infty} \frac{2 \Lambda_n(n^2 \lambda / 2)}{n^2}.$$

Suppose that f is the graphon of an Erdős–Rényi graph G with parameters n and p. Recall the set S of all symmetric L^2 functions, defined in the statement of Lemma 5.2. Take any $a \in S$ and let $\hat{a}_n$ be the level n approximant of a, as defined in Chap. 2, Sect. 2.2. Let $X_{ij} = 1$ if $\{i, j\}$ is an edge in G and 0 otherwise, so that $X_{ij} = X_{ji}$ and $X_{ii} = 0$. Let $B_{i,j,n}$ be the square $[(i-1)/n, i/n] \times [(j-1)/n, j/n]$. Then

$$\lambda_a(f) = \sum_{i,j=1}^{n} X_{ij} \int_{B(i,j,n)} a(x, y) \, dx \, dy$$

$$= \sum_{1 \le i < j \le n} X_{ij} \int_{B(i,j,n) \cup B(j,i,n)} a(x, y) \, dx \, dy.$$

Since the X_{ij}'s are independent and

$$\mathbb{E}(e^{\theta X_{ij}}) = pe^{\theta} + 1 - p$$

for any θ, and a is symmetric, the above expression yields

$$\Lambda_n(n^2 \lambda_a / 2)$$

$$= \log \prod_{1 \le i < j \le n} \left(p \exp\left(\frac{n^2}{2} \int_{B(i,j,n) \cup B(j,i,n)} a(x, y) \, dx \, dy \right) + 1 - p \right)$$

$$= n^2 \sum_{1 \le i < j \le n} \int_{B(i,j,n)} \log(p e^{\hat{a}_n(x,y)} + 1 - p) \, dx \, dy$$

$$= \frac{n^2}{2} \int_{[0,1]^2 \setminus B_n} \log(p e^{\hat{a}_n(x,y)} + 1 - p) \, dx \, dy,$$

where

$$B_n := \bigcup_{i=1}^{n} B(i, i, n).$$

Now note that the function $u(x) = \log(pe^x + 1 - p)$ has derivative $u'(x) = pe^x/(pe^x + 1 - p)$, whose absolute value is bounded everywhere by 1. Therefore $|u(x) - u(y)| \le |x - y|$ for all $x, y \in \mathbb{R}$. Consequently,

$$|\log(pe^{\hat{a}_n(x,y)} + 1 - p) - \log(pe^{a(x,y)} + 1 - p)| \le |\hat{a}_n(x, y) - a(x, y)|.$$

By Proposition 2.6 of Chap. 2, $\hat{a}_n \to a$ in L^2. Therefore the above inequality (together with an application of the Cauchy–Schwarz inequality) implies that

$$\lim_{n \to \infty} \int_{[0,1]^2} \log(pe^{\hat{a}_n(x,y)} + 1 - p) \, dx \, dy = \int_{[0,1]^2} \log(pe^{a(x,y)} + 1 - p) \, dx \, dy.$$

On the other hand, note that

$$|\log(pe^x + 1 - p)| = |u(x) - u(0)| \le |x|.$$

Therefore by another application of the Cauchy–Schwarz inequality,

$$\left| \int_{B_n} \log(pe^{\hat{a}_n(x,y)} + 1 - p) \, dx \, dy \right| \le \left(\mathrm{Leb}(B_n) \int_{B_n} \hat{a}_n(x, y)^2 \, dx \, dy \right)^{1/2}$$

$$\le \|\hat{a}_n\| \sqrt{\mathrm{Leb}(B_n)},$$

where $\mathrm{Leb}(B_n)$ is the Lebesgue measure of B_n and $\|\hat{a}_n\|$ is the L^2 norm of $\hat{a}_n$. Since $\mathrm{Leb}(B_n) = 1/n$ and $\hat{a}_n \to a$ in L^2, this implies that

$$\lim_{n \to \infty} \int_{B_n} \log(pe^{\hat{a}_n(x,y)} + 1 - p) \, dx \, dy = 0.$$

Combining the above observations, we get that for any symmetric $a \in \mathscr{X}$,

$$\bar{\Lambda}(\lambda_a) = \int_{[0,1]^2} \log(pe^{a(x,y)} + 1 - p) \, dx \, dy.$$

For any $f \in \mathscr{X}$, let

$$\bar{\Lambda}^*(f) := \sup_{\lambda \in \mathscr{X}^*} (\lambda(f) - \bar{\Lambda}(\lambda)).$$

Then by Lemma 5.2 and the above calculation, for any $f \in \mathscr{W}$,

$$\bar{\Lambda}^*(f) \geq \sup_{a \in S}(\lambda_a(f) - \bar{\Lambda}(\lambda_a))$$

$$= \sup_{a \in S} J_{p,a}(f) = I_p(f).$$

The proof is now completed by invoking the compactness of the weak topology (Chap. 2, Proposition 2.8) and the abstract large deviation upper bound for topological vector spaces (Chap. 4, Theorem 4.1).   □

5.4   Large Deviation Principle for Dense Random Graphs

Given $p \in (0, 1)$, the following theorem says that the sequence of probability measures $\{\widetilde{\mathbb{P}}_{n,p}\}_{n \geq 1}$ on $\widetilde{\mathscr{W}}$, defined in Sect. 5.1, satisfies a large deviation principle with rate $2/n^2$ and rate function I_p for the cut topology. This is the main result of this chapter. It is much more useful than an LDP for the weak topology, since a large class of functions, including homomorphism densities, are continuous with respect to the cut topology but are not continuous with respect to the weak topology. The word 'dense' in the title of the section refers to the fact that p is fixed and not decaying to zero as n goes to infinity, which means that the number of edges is of the same order as the square of the number of vertices.

Theorem 5.2 *Take any $p \in (0, 1)$, and consider the space $\widetilde{\mathscr{W}}$ equipped with the cut metric $\delta_\square$ defined in Chap. 3. Then for any closed set $\widetilde{F} \subseteq \widetilde{\mathscr{W}}$,*

$$\limsup_{n \to \infty} \frac{2}{n^2} \log \widetilde{\mathbb{P}}_{n,p}(\widetilde{F}) \leq -\inf_{\tilde{h} \in F} I_p(\tilde{h})$$

and for any open set $\widetilde{U} \subseteq \widetilde{\mathscr{W}}$,

$$\liminf_{n \to \infty} \frac{2}{n^2} \log \widetilde{\mathbb{P}}_{n,p}(\widetilde{U}) \geq -\inf_{\tilde{h} \in U} I_p(\tilde{h}).$$

We will first prove the upper bound. The proof will use results that have been proved in the previous three chapters, as well as the weak upper bound proved in the previous section.

For $\tilde{h} \in \widetilde{\mathscr{W}}$ and $\eta > 0$, define

$$S_\square(\tilde{h}, \eta) = \{\tilde{g} \in \widetilde{\mathscr{W}} : \delta_\square(\tilde{g}, \tilde{h}) \leq \eta\}.$$

To prove the upper bound of the theorem, Lemma 4.1 of Chap. 4 says that it suffices to prove that for every $\tilde{h} \in \widetilde{\mathcal{W}}$,

$$\lim_{\eta \to 0} \limsup_{n \to \infty} \frac{2}{n^2} \log \widetilde{\mathbb{P}}_{n,p}(S_\square(\tilde{h}, \eta)) \le -I_p(\tilde{h}). \tag{5.4.1}$$

Let $B(\tilde{h}, \eta) \subset \mathcal{W}$ be defined as

$$B(\tilde{h}, \eta) = \{g \in \mathcal{W} : \tilde{g} \in S_\square(\tilde{h}, \eta)\}.$$

Then by the definition of $\widetilde{\mathbb{P}}_{n,p}$,

$$\mathbb{P}_{n,p}(B(\tilde{h}, \eta)) = \widetilde{\mathbb{P}}_{n,p}(S_\square(\tilde{h}, \eta)).$$

Therefore, by (5.4.1), we only need to show that for every $\tilde{h} \in \widetilde{\mathcal{W}}$,

$$\lim_{\eta \to 0} \limsup_{n \to \infty} \frac{1}{n^2} \log \mathbb{P}_{n,p}(B(\tilde{h}, \eta)) \le -I_p(\tilde{h}). \tag{5.4.2}$$

Now recall the sets $\mathcal{W}(\epsilon)$ from Theorem 3.1 of Chap. 3.

Lemma 5.3 *Take any $\epsilon > 0$ and let $\mathcal{W}(\epsilon)$ be as in Theorem 3.1. Then for any $\tilde{h} \in \widetilde{\mathcal{W}}$ and $\eta > 0$,*

$$\mathbb{P}_{n,p}(B(\tilde{h}, \eta)) \le n!\, \mathbb{P}_{n,p}(B(\tilde{h}, \eta) \cap B(\mathcal{W}(\epsilon), \epsilon)),$$

where $B(\mathcal{W}(\epsilon), \epsilon) = \{g \in \mathcal{W} : \min_{f \in \mathcal{W}(\epsilon)} d_\square(g, f) \le \epsilon\}$.

Proof Let G be a $G(n, p)$ random graph, defined on a probability space $(\Omega, \mathcal{F}, \mathbb{P})$. By Theorem 3.1, there exists $\sigma \in \mathcal{M}_n$ and $h \in \mathcal{W}(\epsilon)$ such that $d_\square(f_\sigma^G, h) < \epsilon$. Thus,

$$\mathbb{P}_{n,p}(B(\tilde{h}, \eta)) = \mathbb{P}(f^G \in B(\tilde{h}, \eta))$$

$$\le \sum_{\sigma \in \mathcal{W}_n} \mathbb{P}(f_\sigma^G \in B(\tilde{h}, \eta) \cap B(\mathcal{W}(\epsilon), \epsilon)).$$

The proof is completed by observing that f_σ^G has the same probability law as f^G for any $\sigma \in \mathcal{M}_n$ and $|\mathcal{M}_n| = n!$.   □

For any $g \in \mathcal{W}$ and $\epsilon > 0$, define

$$B(g, \epsilon) := \{h \in \mathcal{W} : d_\square(h, g) \le \epsilon\}.$$

The following lemma provides an important bridge between the weak and cut topologies.

Lemma 5.4 *For any $g \in \mathcal{W}$ and $\epsilon > 0$, $B(g, \epsilon)$ is weakly closed.*

Proof Suppose that $\{g_n\}_{n\geq 1}$ is a sequence in $\mathscr{W}$ such that $g_n \in B(g, \epsilon)$ for each n and $g_n \to h$ weakly. Take any two Borel measurable functions $a, b : [0, 1] \to [-1, 1]$. Since $g_n \to h$ weakly,

$$\left| \int_{[0,1]^2} a(x)b(y)(h(x, y) - g(x, y)) \, dx \, dy \right|$$

$$= \lim_{n \to \infty} \left| \int_{[0,1]^2} a(x)b(y)(g_n(x, y) - g(x, y)) \, dx \, dy \right| \leq \epsilon.$$

Taking supremum over all a and b gives $d_\square(h, g) \leq \epsilon$. $\square$

Lemma 5.5 *There exists a function* $\delta(\tilde{h}, \epsilon)$, *depending only on* $\tilde{h}$ *and* ϵ, *with* $\delta(\tilde{h}, \epsilon) \to 0$ *as* $\epsilon \to 0$, *such that for each* $\tilde{h} \in \widetilde{\mathscr{W}}$, $\eta > 0$ *and* $\epsilon > 0$,

$$\lim_{\eta \to 0} \limsup_{n \to \infty} \frac{1}{n^2} \log \mathbb{P}_{n,p}(B(\tilde{h}, \eta) \cap B(\mathscr{W}(\epsilon), \epsilon)) \leq -I_p(\tilde{h}) + \delta(\tilde{h}, \epsilon).$$

Proof Since $\mathscr{W}(\epsilon)$ is a finite set, it suffices to show that for fixed $g \in \mathscr{W}(\epsilon)$,

$$\lim_{\eta \to 0} \limsup_{n \to \infty} \frac{1}{n^2} \log \mathbb{P}_{n,p}(B(\tilde{h}, \eta) \cap B(g, \epsilon)) \leq -I_p(\tilde{h}) + \delta(\tilde{h}, \epsilon).$$

If $B(\tilde{h}, \eta) \cap B(g, \epsilon)$ is empty for sufficiently small η, then there is nothing to prove. So let us assume that this is not the case. Then

$$g \in B(\tilde{h}, \epsilon). \tag{5.4.3}$$

By lower semi-continuity of I_p, $I_p(f) \geq I_p(\tilde{h}) - \delta(\tilde{h}, \epsilon)$ for $f \in B(\tilde{h}, 2\epsilon)$, where $\delta(\tilde{h}, \epsilon) \to 0$ as $\epsilon \to 0$. But by (5.4.3), $B(g, \epsilon) \subseteq B(\tilde{h}, 2\epsilon)$ and by Lemma 5.4, $B(g, \epsilon)$ is weakly closed. Therefore by Theorem 5.1,

$$\lim_{\eta \to 0} \limsup_{n \to \infty} \frac{1}{n^2} \log \mathbb{P}_{n,p}(B(\tilde{h}, \eta) \cap B(g, \epsilon))$$

$$\leq \limsup_{n \to \infty} \frac{1}{n^2} \log \mathbb{P}_{n,p}(B(g, \epsilon))$$

$$\leq - \inf_{f \in B(g,\epsilon)} I_p(f)$$

$$\leq - \inf_{f \in B(\tilde{h},2\epsilon)} I_p(f) \leq -I_p(\tilde{h}) + \delta(\tilde{h}, \epsilon).$$

This completes the proof of the lemma. $\square$

We are now ready to prove the upper bound of Theorem 5.2.

Proof (Proof of the Upper Bound in Theorem 5.2) The combination of Lemma 5.3 and Lemma 5.5 gives the inequality (5.4.2) after taking $\epsilon \to 0$, which completes the proof of the upper bound of Theorem 5.2. □

Let us now turn our attention to the lower bound of Theorem 5.2. As for the upper bound, we will first go through a sequence of reductions. First, note that for the lower bound it suffices to prove that for all $\tilde{h} \in \widetilde{\mathcal{W}}$ and $\eta \in (0, 1)$,

$$\liminf_{n \to \infty} \frac{1}{n^2} \log \widetilde{\mathbb{P}}_{n,p}(S_{\square}(\tilde{h}, \eta)) \geq -I_p(\tilde{h}),$$

since for any open $\widetilde{U}$ and $\tilde{h} \in \widetilde{U}$, there exists $\eta \in (0, 1)$ such that $S_{\square}(\tilde{h}, \eta) \subseteq \widetilde{U}$. Again, since $B(h, \eta) \subseteq B(\tilde{h}, \eta)$ and $\mathbb{P}_{n,p}(B(\tilde{h}, \eta)) = \widetilde{\mathbb{P}}_{n,p}(S_{\square}(\tilde{h}, \eta))$, it suffices to prove that for any $h \in \mathcal{W}$ and $\eta \in (0, 1)$,

$$\liminf_{n \to \infty} \frac{1}{n^2} \log \mathbb{P}_{n,p}(B(h, \eta)) \geq -I_p(h).$$

Take any $h \in \mathcal{W}$ and $\eta > 0$. Let $\hat{h}_n$ be the level n approximant of h, as defined in Chap. 2, Sect. 2.2. Fix $\epsilon \in (0, \eta)$. By Proposition 2.6, $\hat{h}_n \to h$ in L^2. An easy application of the Cauchy–Schwarz inequality shows that the L^2 topology is finer than the cut topology. Therefore $\hat{h}_n \to h$ in the cut metric. Consequently, $B(\hat{h}_n, \epsilon) \subseteq B(h, \eta)$ for all large n. Thus, it suffices to prove that

$$\liminf_{n \to \infty} \frac{1}{n^2} \log \mathbb{P}_{n,p}(B(\hat{h}_n, \epsilon)) \geq -I_p(h).$$

As in the proof of Theorem 5.1, let $B(i, j, n)$ denote the square $[(i-1)/n, i/n] \times [(j-1)/n, j/n]$ and

$$B_n = \bigcup_{i=1}^{n} B(i, i, n).$$

Note that $\hat{h}_n$ is constant in any such square (except possibly at the boundaries). Define a function q_n that equals $\hat{h}_n$ everywhere except on B_n, where it is zero. Since the Lebesgue measure of B_n tends to zero as $n \to \infty$ and q_n and $\hat{h}_n$ are uniformly bounded functions, $\hat{h}_n - q_n \to 0$ in L^2 and so $d_{\square}(\hat{h}_n, q_n) \to 0$. Therefore, it suffices to prove that for any $\epsilon \in (0, 1)$,

$$\liminf_{n \to \infty} \frac{1}{n^2} \log \mathbb{P}_{n,p}(B(q_n, \epsilon)) \geq -I_p(h). \tag{5.4.4}$$

Let $q(i, j, n)$ be the value of q_n in $B(i, j, n)$. Construct a random graph H_n on n vertices (on some abstract probability space $(\Omega, \mathscr{F}, \mathbb{P})$) by declaring that vertices i and j are connected by an edge with probability $q(i, j, n)$, for every $1 \leq i < j \leq n$.

Let $\xi(i, j, n)$ be a random variable that is 1 if the edge $\{i, j\}$ is present in H_n and 0 otherwise. Let f_n denote the graphon of H_n, and let $\mathbb{P}_{n,h}$ denote the law of f_n.

Lemma 5.6 *Let f_n and q_n be as above. Then for any Borel measurable a, b : $[0, 1] \to [-1, 1]$ and any $x \geq 0$,*

$$\mathbb{P}\left(\left|\int_{[0,1]^2} a(x)b(y)(f_n(x, y) - q_n(x, y))\, dx\, dy\right| \geq x\right) \leq 2e^{-n^2 x^2/4}.$$

Proof For $1 \leq i \leq n$, let

$$a_i := n \int_{(i-1)/n}^{i/n} a(x)\, dx,$$

and define b_i similarly. Then

$$\int_{[0,1]^2} a(x)b(y)(f_n(x, y) - q_n(x, y))\, dx\, dy$$

$$= \frac{2}{n^2} \sum_{1 \leq i < j \leq n} a_i b_j (\xi(i, j, n) - q(i, j, n)).$$

Since this is an average of independent bounded random variables with mean zero, it is now easy to complete the proof using the Azuma–Hoeffding inequality (Theorem 4.2).   □

The support of the measure $\mathbb{P}_{n,h}$ is contained in the support of $\mathbb{P}_{n,p}$, which consists of all graphons of simple graphs on n vertices. Take any such graphon f, and let f_{ij} denote its value in the interior of $B(i, j, n)$. Then

$$\mathbb{P}_{n,h}(\{f\}) = \prod_{1 \leq i < j \leq n} q(i, j, n)^{f_{ij}} (1 - q(i, j, n))^{1-f_{ij}}$$

and

$$\mathbb{P}_{n,p}(\{f\}) = \prod_{1 \leq i < j \leq n} p^{f_{ij}} (1 - p)^{1-f_{ij}}.$$

These formulas yield the following lemma.

Lemma 5.7 *Let $\mathbb{P}_{n,h}$ be defined as above. Then*

$$\lim_{n \to \infty} \frac{2}{n^2} \int_{\mathscr{W}} \log \frac{d\mathbb{P}_{n,h}}{d\mathbb{P}_{n,p}}\, d\mathbb{P}_{n,h} = I_p(h).$$

Proof From the formulas for $\mathbb{P}_{n,h}$ and $\mathbb{P}_{n,p}$ displayed above, it follows that

$$\frac{2}{n^2} \int_{\mathscr{W}} \log \frac{d\mathbb{P}_{n,h}}{d\mathbb{P}_{n,p}}(f) \, d\mathbb{P}_{n,h}(f)$$

$$= \frac{2}{n^2} \sum_{1 \le i < j \le n} \int_{\mathscr{W}} \left(f_{ij} \log \frac{q(i,j,n)}{p} + (1 - f_{ij}) \log \frac{1 - q(i,j,n)}{1 - p} \right) d\mathbb{P}_{n,h}(f)$$

$$= \frac{2}{n^2} \sum_{1 \le i < j \le n} \left(q(i,j,n) \log \frac{q(i,j,n)}{p} + (1 - q(i,j,n)) \log \frac{1 - q(i,j,n)}{1 - p} \right)$$

$$= I_p(q_n).$$

Next, take any $\epsilon > 0$ and let

$$A_{n,\epsilon} := \{ (x,y) \in [0,1]^2 : |q_n(x,y) - h(x,y)| \ge \epsilon \}.$$

By Chebychev's inequality (Lemma 2.1),

$$\mathrm{Leb}(A_{n,\epsilon}) \le \frac{\|q_n - h\|^2}{\epsilon^2},$$

where $\mathrm{Leb}(A_{n,\epsilon})$ is the Lebesgue measure of $A_{n,\epsilon}$ and $\|q_n - h\|$ is the L^2 norm of $q_n - h$. Now note that I_p is bounded and uniformly continuous on $[0,1]$. If $I_p(x) \le C(p)$ for all x and $|I_p(x) - I_p(y)| < \delta(\epsilon)$ whenever $|x - y| < \epsilon$, the above inequality shows that

$$|I_p(q_n) - I_p(h)| \le \int_{A_{n,\epsilon}} |I_p(q_n(x,y)) - I_p(h(x,y))| \, dx \, dy$$

$$+ \int_{[0,1]^2 \setminus A_{n,\epsilon}} |I_p(q_n(x,y)) - I_p(h(x,y))| \, dx \, dy$$

$$\le C(p) \mathrm{Leb}(A_{n,\epsilon}) + \delta(\epsilon)$$

$$\le \frac{C(p) \|q_n - h\|^2}{\epsilon^2} + \delta(\epsilon).$$

The proof is completed by letting $n \to \infty$ and then sending $\epsilon \to 0$.   □

Recall the sets $\mathscr{W}_n$ and $\mathscr{U}_n$ defined in Chap. 3, Sect. 3.5. Let $\mathscr{U}_n'$ be the subset of $\mathscr{U}_n$ consisting of all functions whose ranges are contained in $[-1,1]$.

Lemma 5.8 *If $f, g \in \mathscr{W}_n$, then*

$$d_\square(f,g) = \sup_{a,b \in \mathscr{U}_n'} \left| \int_{[0,1]^2} a(x) b(y) (f(x,y) - g(x,y)) \, dx \, dy \right|.$$

Proof Take any Borel measurable $a, b : [0, 1] \to [-1, 1]$. Let

$$\bar{a}(x) = a_i := n \int_{(i-1)/n}^{i/n} a(y)\, dy \quad \text{if } \frac{i-1}{n} \le x < \frac{i}{n},$$

for $i = 1, \ldots, n$, and let $\bar{a}(1) = 0$. Define $\bar{b}$ similarly. Then $\bar{a}, \bar{b} \in \mathcal{U}_n'$. Let $B(i, j, n)$ be as in the proof of Theorem 5.1. Let f_{ij} be the value of in the interior of $B(i, j, n)$, and define g_{ij} similarly. Then

$$\int_{[0,1]^2} a(x)b(y)(f(x, y) - g(x, y))\, dx\, dy$$

$$= \sum_{i,j=1}^{n} (f_{ij} - g_{ij}) \int_{B(i,j,n)} a(x)b(y)\, dx\, dy$$

$$= \frac{1}{n^2} \sum_{i,j=1}^{n} (f_{ij} - g_{ij}) a_i b_j$$

$$= \int_{[0,1]^2} \bar{a}(x)\bar{b}(y)(f(x, y) - g(x, y))\, dx\, dy.$$

Therefore, in the computation of $d_\square(f, g)$, it suffices to maximize over $\mathcal{U}_n'$ instead of all Borel measurable a, b. $\square$

Lemma 5.9 *Given any $\epsilon \in (0, 1)$, there exists a set $\mathcal{U}_n'(\epsilon) \subseteq \mathcal{U}_n'$ of size $\le (3/\epsilon)^n$ such that for any $a \in \mathcal{U}_n'$, there exists $b \in \mathcal{U}_n'(\epsilon)$ with $|a(x) - b(x)| \le \epsilon$ for every $x \in [0, 1]$ that is not of the form i/n for some integer $0 \le i \le n$.*

Proof Take any i. By the definition of $\mathcal{U}_n'$, the value of a is constant in the interval $((i - 1)/n, i/n)$. In this interval, define $b(x)$ to be $k\epsilon$, where k is an integer chosen such that this number is closest to the value of a and $k\epsilon \in [-1, 1]$. The number of possible values of k is at most $2/\epsilon + 1 \le 3/\epsilon$. $\square$

Let $\mathcal{U}_n'(\epsilon)$ be as in the previous lemma. For $\epsilon > 0$ and $f, g \in \mathcal{W}_n$,

$$d_\square^\epsilon(f, g) := \max_{a,b \in \mathcal{U}_n'(\epsilon)} \left| \int_{[0,1]^2} a(x)b(y)(f(x, y) - g(x, y))\, dx\, dy \right|.$$

The following lemma shows how $d_\square(f, g)$ may be approximated by $d_\square^\epsilon(f, g)$.

Lemma 5.10 *For any $f, g \in \mathcal{W}_n$ and $\epsilon \in (0, 1/2)$,*

$$d_\square(f, g) \le \frac{d_\square^\epsilon(f, g)}{1 - 2\epsilon}.$$

Proof Take any $a, b \in \mathcal{U}_n'$. Find $a', b' \in \mathcal{U}_n'(\epsilon)$ satisfying the approximation property of Lemma 5.9. Then there exists $c, d : [0, 1] \to [-\epsilon, \epsilon]$ such that $a - a' = c$

and $b - b' = d$ almost everywhere. Therefore by the definition of the cut metric,

$$\left| \int_{[0,1]^2} (a(x)b(y) - a'(x)b'(y))(f(x,y) - g(x,y))\,dx\,dy \right|$$

$$\leq \left| \int_{[0,1]^2} c(x)b(y)(f(x,y) - g(x,y))\,dx\,dy \right|$$

$$+ \left| \int_{[0,1]^2} a'(x)d(y)(f(x,y) - g(x,y))\,dx\,dy \right|$$

$$\leq 2\epsilon d_\square(f,g).$$

Therefore by Lemma 5.8,

$$d_\square(f,g) \leq d_\square^\epsilon(f,g) + 2\epsilon d_\square(f,g).$$

This completes the proof. $\square$

Lemma 5.11 *For any $\epsilon \in (0,1)$,*

$$\lim_{n\to\infty} \mathbb{P}_{n,h}(B(q_n, \epsilon)) = 1.$$

Proof Fix $\epsilon \in (0,1)$. Let f_n be as in Lemma 5.6. Then by Lemma 5.10,

$$1 - \mathbb{P}_{n,h}(B(q_n, \epsilon)) = \mathbb{P}(d_\square(f_n, q_n) > \epsilon) \leq \mathbb{P}(d_\square^{1/4}(f_n, q_n) > \epsilon/2).$$

Applying Lemma 5.6 with $a, b \in \mathcal{U}_n'(1/4)$ and $x = \epsilon/2$, we get

$$\mathbb{P}(d_\square^{1/4}(f_n, q_n) > \epsilon/2) \leq 2|\mathcal{U}_n'(1/4)|e^{-n^2\epsilon^2/16}.$$

The bound on $|\mathcal{U}_n'(1/4)|$ from Lemma 5.9 shows that the last expression tends to zero as $n \to \infty$. $\square$

We are now ready to prove the lower bound of Theorem 5.2. We will continue to use the notations introduced in the last few pages.

Proof (Proof of the Lower Bound in Theorem 5.2) Note that

$$\mathbb{P}_{n,p}(B(q_n, \epsilon)) = \int_{B(q_n,\epsilon)} d\mathbb{P}_{n,p} = \int_{B(q_n,\epsilon)} \exp\left(-\log \frac{d\mathbb{P}_{n,h}}{d\mathbb{P}_{n,p}}\right) d\mathbb{P}_{n,h}$$

$$= \mathbb{P}_{n,h}(B(q_n, \epsilon)) \frac{1}{\mathbb{P}_{n,h}(B(q_n, \epsilon))} \int_{B(q_n,\epsilon)} \exp\left(-\log \frac{d\mathbb{P}_{n,h}}{d\mathbb{P}_{n,p}}\right) d\mathbb{P}_{n,h}.$$

Therefore, by Jensen's inequality (Proposition 2.1),

$$\log \mathbb{P}_{n,p}(B(q_n, \epsilon)) \geq \log \mathbb{P}_{n,h}(B(q_n, \epsilon))$$

$$- \frac{1}{\mathbb{P}_{n,h}(B(q_n, \epsilon))} \int_{B(q_n,\epsilon)} \log \frac{d\mathbb{P}_{n,h}}{d\mathbb{P}_{n,p}} \, d\mathbb{P}_{n,h}.$$

Since $\mathbb{P}_{n,h}(B(q_n, \epsilon)) \to 1$ by Lemma 5.11, this implies that

$$\liminf_{n\to\infty} \frac{2}{n^2} \log \mathbb{P}_{n,p}(B(q_n, \epsilon)) \geq - \lim_{n\to\infty} \frac{2}{n^2} \int \log \frac{d\mathbb{P}_{n,h}}{d\mathbb{P}_{n,p}} \, d\mathbb{P}_{n,h}.$$

By Lemma 5.7, the expression on the right equals $-I_p(h)$. This proves (5.4.4) and hence the lower bound of Theorem 5.2. □

5.5 Conditional Distributions

Let G be a $G(n, p)$ random graph. Let f^G be the graphon of G and let $\widetilde{f^G} \in \widetilde{\mathcal{W}}$ be the equivalence class of this graphon. In this section we will denote $\widetilde{f^G}$ simply by G, for ease of notation. Theorem 5.2 gives estimates of the probabilities of rare events for G. However, it does not answer the following question: given that some particular rare event has occurred, what does the graph look like? Naturally, one might expect that if $G \in \widetilde{F}$ for some closed set $\widetilde{F} \subseteq \widetilde{\mathcal{W}}$ satisfying

$$\inf_{\tilde{h} \in \widetilde{F}^o} I_p(\tilde{h}) = \inf_{\tilde{h} \in \widetilde{F}} I_p(\tilde{h}) > 0, \tag{5.5.1}$$

then G should resemble one of the minimizers of I_p in $\widetilde{F}$. (Here $\widetilde{F}^o$ denotes the interior of $\widetilde{F}$.) In other words, given that $G \in \widetilde{F}$, one might expect that $\delta_\square(G, \widetilde{F}^*) \approx 0$, where $\widetilde{F}^*$ is the set of minimizers of I_p in $\widetilde{F}$ and

$$\delta_\square(G, \widetilde{F}^*) := \inf_{\tilde{h} \in \widetilde{F}^*} \delta_\square(G, \tilde{h}). \tag{5.5.2}$$

However, it is not obvious that a minimizer must exist in $\widetilde{F}$. The compactness of $\widetilde{\mathcal{W}}$ comes to the rescue: since the function I_p is lower semicontinuous on $\widetilde{F}$ and $\widetilde{F}$ is closed, a minimizer must necessarily exist. The following theorem formalizes this argument.

Theorem 5.3 *Take any $p \in (0, 1)$ and $n \geq 1$. Let G be a random graph from the $G(n, p)$ model. Let $\widetilde{F}$ be a closed subset of $\widetilde{\mathcal{W}}$ satisfying (5.5.1). Let $\widetilde{F}^*$ be the subset of $\widetilde{F}$ where I_p is minimized. Then $\widetilde{F}^*$ is non-empty and compact, and for each n, and* *each $\epsilon > 0$,*

$$\mathbb{P}(\delta_\square(G, \widetilde{F}^*) \geq \epsilon \mid G \in \widetilde{F}) \leq e^{-C(\epsilon, \widetilde{F})n^2}$$

where $C(\epsilon, \widetilde{F})$ is a positive constant depending only on ϵ and $\widetilde{F}$ and $\delta_\square(G, \widetilde{F}^*)$ is defined as in (5.5.2). In particular, if $\widetilde{F}^*$ contains only one element $\tilde{h}^*$, then the conditional distribution of G given $G \in \widetilde{F}$ converges to the point mass at $\tilde{h}^*$ as $n \to \infty$.

Proof Since $\widetilde{\mathscr{W}}$ is compact and $\widetilde{F}$ is a closed subset, $\widetilde{F}$ is also compact. Since I_p is a lower semicontinuous function on $\widetilde{F}$ (Proposition 5.1) and $\widetilde{F}$ is compact, it must attain its minimum on $\widetilde{F}$. Thus, $\widetilde{F}^*$ is non-empty. By the lower semicontinuity of I_p, $\widetilde{F}^*$ is closed (and hence compact). Fix $\epsilon > 0$ and let

$$\widetilde{F}_\epsilon := \{\tilde{h} \in \widetilde{F} : \delta_\square(\tilde{h}, \widetilde{F}^*) \geq \epsilon\}.$$

Then $\widetilde{F}_\epsilon$ is again a closed subset. Observe that

$$\mathbb{P}(\delta_\square(G, \widetilde{F}^*) \geq \epsilon \mid G \in \widetilde{F}) = \frac{\mathbb{P}(G \in \widetilde{F}_\epsilon)}{\mathbb{P}(G \in \widetilde{F})}.$$

Thus, with

$$I_1 := \inf_{\tilde{h} \in \widetilde{F}} I_p(\tilde{h}), \quad I_2 := \inf_{\tilde{h} \in \widetilde{F}_\epsilon} I_p(\tilde{h}),$$

Theorem 5.2 and condition (5.5.1) give

$$\limsup_{n \to \infty} \frac{1}{n^2} \log \mathbb{P}(\delta_\square(G, \widetilde{F}^*) \geq \epsilon \mid G \in \widetilde{F}) \leq I_1 - I_2.$$

The proof will be complete if it is shown that $I_1 < I_2$. Now clearly, $I_1 \leq I_2$. If $I_1 = I_2$, the compactness of $\widetilde{F}_\epsilon$ implies that there exists $\tilde{h} \in \widetilde{F}_\epsilon$ satisfying $I_p(\tilde{h}) = I_2$. However, this means that $\tilde{h} \in \widetilde{F}^*$ and hence $\widetilde{F}_\epsilon \cap \widetilde{F}^* \neq \emptyset$, which is impossible. $\square$

Bibliographical Notes

The Erdős–Rényi model was introduced by Gilbert [5] and Erdős and Rényi [3]. It has been the subject of extensive investigations over the years. See Bollobás [1] and Janson et al. [6] for partial surveys of this literature.

The key feature of the Erdős–Rényi model, which makes it amenable to a lot of beautiful mathematics, is that the edges are independent. In the large deviation regime, however, the independence is lost: conditional on a rare event, the edges typically stop behaving as independent random objects. This is the main difficulty behind the large deviation analysis of the Erdős–Rényi model, which remained open for many years. The results of this chapter were proved by Chatterjee and Varadhan [2], generalizing a conjecture of Bolthausen et al. (Large deviations for random matrices and random graphs, Private communication, 2003) concerning large deviations for subgraph counts. The original proof of Theorem 5.2, as it appeared in Chatterjee and Varadhan [2], was based on Szemerédi's regularity lemma. The proof given in this chapter uses Theorem 3.1, which can be called a version of the weak regularity theorem of Frieze and Kannan [4], adapted to the setting of graphons. While the published proof of Theorem 5.2 was combinatorial in nature, the proof given in this monograph is analytic.

References

1. Bollobás, B. (2001). *Random graphs*, 2nd ed., vol. 73, Cambridge studies in advanced mathematics. Cambridge: Cambridge University Press.
2. Chatterjee, S., & Varadhan, S. R. S. (2011). The large deviation principle for the Erdős–Rényi random graph. *European Journal of Combinatorics, 32*(7), 1000–1017.
3. Erdős, P., & Rényi, A. (1960). On the evolution of random graphs. *Publication of the Mathematical Institute of the Hungarian Academy of Sciences, 5*, 17–61.
4. Frieze, A., & Kannan, R. (1999). Quick approximation to matrices and applications. *Combinatorica, 19*, 175–220.
5. Gilbert, E. N. (1959). Random graphs. *Annals of Mathematical Statistics, 30*(4), 1141–1144.
6. Janson, S., Łuczak, T., & Ruciński, A. (2000). *Random graphs*. Wiley-interscience series in discrete mathematics and optimization. New York: Wiley-Interscience.

Chapter 6
Applications of Dense Graph Large Deviations

This chapter contains some simple applications of the large deviation principle for dense Erdős–Rényi random graphs that was derived in the previous chapter. The abstract theory yields surprising phase transition phenomena when applied to concrete problems. We will continue to use notations and terminologies introduced previously in the monograph.

6.1 Graph Parameters

A graph parameter is a continuous function from $\widetilde{\mathscr{W}}$ into $\mathbb{R}$. Any graph parameter has a natural interpretation as a continuous function on $\mathscr{W}$. We will generally use the same letter to denote a graph parameter and its lift to $\mathscr{W}$.

The L^∞ norm of a graphon f, denoted by $\|f\|_\infty$, is the essential supremum of $|f(x,y)|$ as (x,y) ranges over $[0,1]^2$. A graphon $f \in \mathscr{W}$ is called a local maximum with respect to the L^∞ norm for a graph parameter τ if there exists $\epsilon > 0$ such that for any $g \in \mathscr{W}$ with $\|f - g\|_\infty \leq \epsilon$, we have $\tau(g) \leq \tau(f)$. Local minimum is defined similarly. A graphon f is called a global maximum if $\tau(f) \geq \tau(g)$ for all g, and a global minimum if $\tau(f) \leq \tau(g)$ for all g.

The following special kind of graph parameters are important in this chapter.

Definition 6.1 A graph parameter τ will be called a 'nice graph parameter' if every local maximum of τ with respect to the L^∞ norm is a global maximum and every local minimum of τ with respect to the L^∞ norm is a global minimum.

The next two lemmas identify two examples of nice graph parameters. For the first, recall the definition of the homomorphism density $t(H, f)$ from Chap. 3.

Lemma 6.1 *For any simple graph H with at least one edge, the function $\tau(f) := t(H, f)$ is a nice graph parameter.*

© Springer International Publishing AG 2017
S. Chatterjee, *Large Deviations for Random Graphs*, Lecture Notes in Mathematics 2197, DOI 10.1007/978-3-319-65816-2_6

Proof Combining Proposition 3.2 and Exercise 3.4 from Chap. 3, it follows that τ is continuous with respect to $\delta_\square$ metric. In other words, τ is a graph parameter.

Let f be a local maximum of τ. Choose some $\epsilon > 0$ and let $g^+ := \min\{f+\epsilon, 1\}$. Unless $t(H,f) = 1$, it is easy to see that $t(H, g^+) > t(H,f)$. On the other hand, $\|g-f\|_\infty \leq \epsilon$. This violates the assumption that f is a local maximum of τ. Therefore $t(H,f)$ must be 1. In other words, f must be a global maximum of τ.

Next, suppose that f is a local minimum of τ. Choose some $\epsilon > 0$ and let $g^- := (1 - \epsilon)f$. Unless $t(H,f) = 0$, $t(H, g^-)$ must be strictly less than $t(H,f)$. Arguing as in the previous case, it follows that f must be a global minimum. Thus, τ is a nice graph parameter. $\qquad\square$

For the next lemma, recall the definition of the operator K_f associated with a graphon f, as defined in Sect. 3.4 of Chap. 3. Recall also the definition of the operator norm $\|K_f\|$ of the operator K_f.

Exercise 6.1 If σ is a measure preserving bijection of $[0, 1]$ and $f \in \mathcal{W}$, show that $\|K_f\| = \|K_{f_\sigma}\|$.

Exercise 6.2 For a simple graph G on n vertices, let $\lambda_1(G)$ be the largest eigenvalue of the adjacency matrix of G. Show that $\|K_f\| = \lambda_1(G)/n$. (Hint: Use the Perron–Frobenius theorem, for example the version presented as Proposition 2.11 in this monograph.)

The following lemma shows that the operator norm is a nice graph parameter.

Lemma 6.2 *Let* $\tau(f) := \|K_f\|$. *Then* τ *is a nice graph parameter.*

Proof Let g and h be two graphons and let $f := g - h$. Take any $u \in L^2([0, 1])$ with $\|u\| = 1$. Then by the Cauchy–Schwarz inequality,

$$\|K_f u\|^4 = \left(\int_0^1 \left(\int_0^1 f(x, y)u(y)\, dy\right)^2 dx\right)^2$$

$$= \left(\int_{[0,1]^3} f(x, y)f(x, y')u(y)u(y')\, dx\, dy\, dy'\right)^2$$

$$\leq \left(\int_{[0,1]^2}\left(\int_0^1 f(x, y)f(x, y')\, dx\right)^2 dy\, dy'\right)\left(\int_{[0,1]^2} u(y)^2 u(y')^2\, dy\, dy'\right)$$

$$= \int_{[0,1]^4} f(x, y)f(x, y')f(x', y)f(x', y')\, dx\, dx'\, dy\, dy'.$$

Note that $|f| \leq 1$ everywhere. Thus, for any x', y', the definition of the cut distance implies that

$$\left|\int_{[0,1]^2} f(x, y)f(x, y')f(x', y)\, dx\, dy\right| \leq d_\square(g, h),$$

and so the last expression in the previous display is bounded by $d_\square(g, h)$. Thus, for any $g, h \in \mathscr{W}$,

$$|\tau(g) - \tau(h)|^4 \leq \|K_g - K_h\|^4 \leq d_\square(g, h).$$

Together with Exercise 6.1, this inequality shows that τ is a graph parameter. To show that τ is nice, first observe that there cannot exist any local minima, because for any $\epsilon \in (0, 1)$ and $f \in \mathscr{W}$,

$$\|(1 - \epsilon)K_f\| = (1 - \epsilon)\|K_f\| < \|K_f\|$$

unless $\tau(f) = \|K_f\| = 0$. Next, take any $f \in \mathscr{W}$ and $\epsilon > 0$, and let $g := f + \epsilon(1 - f)$. Then $g \in \mathscr{W}$ and $\|g - f\|_\infty \leq \epsilon$. By Proposition 3.4, K_f is a compact operator. Moreover, K_f is obviously a nonnegative operator, according to the definition of nonnegativity given in the paragraph preceding Proposition 2.11 in Chap. 2. Therefore by Proposition 2.11, there exists $u \in B_1([0, 1])$ such that $K_f u = \|K_f\|u$. Since

$$K_g u = (\|K_f\| + \epsilon(1 - \|K_f\|))u,$$

it follows that $\|K_g\| > \|K_f\|$ unless $\|K_f\| = 1$. Thus, any local maximum of τ must be a global maximum. $\square$

6.2 Rate Functions for Graph Parameters

Recall the function I_p defined in Sect. 5.2 of Chap. 5. Let τ be a nice graph parameter, as defined in the previous section. For any $p \in (0, 1)$ and $t \in \tau(\mathscr{W})$, define

$$\phi_\tau^+(p, t) := \inf\{I_p(f) : f \in \mathscr{W}, \tau(f) \geq t\},$$
$$\phi_\tau^-(p, t) := \inf\{I_p(f) : f \in \mathscr{W}, \tau(f) \leq t\}.$$

The following result establishes an important property of nice graph parameters.

Proposition 6.1 *If τ is a nice graph parameter, then for any $p \in (0, 1)$, the map $t \mapsto \phi_\tau^+(p, t)$ is non-decreasing and continuous and the map $t \mapsto \phi_\tau^-(p, t)$ is non-increasing and continuous.*

Proof Since $-\tau$ is also a nice graph parameter, it suffices to prove the result for ϕ_τ^+. Fix $p \in (0, 1)$. By definition, $t \mapsto \phi_\tau^+(p, t)$ is non-decreasing. Take any $t \in \tau(\mathscr{W})$ and any f such that $\tau(f) \geq t$. First, suppose that $t < \sup \tau(\mathscr{W})$ and let $\{t_n\}_{n \geq 1}$ be a sequence that is strictly decreasing to t. Since every local maximum of τ with respect to the L^∞ norm is a global maximum, there exists a sequence of graphons $\{f_n\}_{n \geq 1}$ such that $\|f_n - f\|_\infty \to 0$ and $\tau(f_n) \geq t_n$ for each n. By the uniform

continuity of the function I_p on $[0, 1]$ and the dominated convergence theorem, it
follows that $I_p(f_n) \to I_p(f)$. Thus,

$$\lim_{n\to\infty} \phi_\tau^+(p, t_n) \leq \lim_{n\to\infty} I_p(f_n) = I_p(f).$$

Since this is true for every f such that $\tau(f) \geq t$ and ϕ_τ^+ is non-decreasing in t, this
proves the right continuity of ϕ_τ^+.

Next, assume that $t > \inf \tau(\mathcal{W})$ and take a sequence $\{t_n\}_{n\geq1}$ that is strictly
increasing to t. Let $\{f_n\}_{n\geq1}$ be a sequence of graphons such that $\tau(f_n) \geq t_n$ for each n
and $\phi_\tau^+(p, t_n) - I_p(f_n) \to 0$. By the invariance of τ and I_p under measure preserving
bijections and the compactness of $\widetilde{\mathcal{W}}$ (Theorem 3.1), we may assume without loss
of generality that there exists a graphon f such that $d_\square(f_n, f) \to 0$. By the continuity
of τ, $\tau(f) \geq t$. By the lower semi-continuity of I_p on $\mathcal{W}$ (Corollary 5.1),

$$\liminf_{n\to\infty} I_p(f_n) \geq I_p(f).$$

Thus,

$$\liminf_{n\to\infty} \phi_\tau^+(p, t_n) \geq I_p(f) \geq \phi_\tau^+(p, t).$$

Since ϕ_τ^+ is non-decreasing, this proves that ϕ_τ^+ is left continuous.   □

6.3   Large Deviations for Graph Parameters

Let τ be a nice graph parameter, and let ϕ_τ^+ and ϕ_τ^- be defined as in the previous
section. For a simple graph G on a finite set of vertices, we will use the notation
$\tau(G)$ to denote $\tau(f^G)$. The following theorem is the main result of this section. It
gives the large deviation rate functions for the upper and lower tails of nice graph
parameters.

Theorem 6.1 *Fix $p \in (0, 1)$. For each n, let $G_{n,p}$ be a random graph from the
$G(n, p)$ model. Let τ be a nice graph parameter. Then for any $t \in \tau(\mathcal{W})$,*

$$\lim_{n\to\infty} \frac{2}{n^2} \log \mathbb{P}(\tau(G_{n,p}) \geq t) = -\phi_\tau^+(p, t),$$

$$\lim_{n\to\infty} \frac{2}{n^2} \log \mathbb{P}(\tau(G_{n,p}) \leq t) = -\phi_\tau^-(p, t).$$

Proof Since $-\tau$ is also a nice graph parameter, it suffices to prove the first identity.
From Theorem 5.2 and the continuity of τ on $\mathcal{W}$, it follows that

$$\limsup_{n\to\infty} \frac{2}{n^2} \log \mathbb{P}(\tau(G_{n,p}) \geq t) \leq -\phi_\tau^+(p, t).$$

Next, let

$$\widetilde{U} := \{\tilde{h} \in \widetilde{\mathscr{W}} : \tau(\tilde{h}) > t\}.$$

By the continuity of τ, $\widetilde{U}$ is an open set. Therefore by Theorem 5.2,

$$\liminf_{n \to \infty} \frac{2}{n^2} \log \mathbb{P}(\tau(G_{n,p}) \geq t) \geq \liminf_{n \to \infty} \frac{2}{n^2} \log \widetilde{\mathbb{P}}_{n,p}(\widetilde{U})$$

$$\geq -\inf_{\tilde{h} \in \widetilde{U}} I_p(\tilde{h}).$$

But it is easy to see that for every $\epsilon > 0$,

$$\inf_{\tilde{h} \in \widetilde{U}} I_p(\tilde{h}) \leq \phi_\tau^+(p, t + \epsilon).$$

By the continuity of ϕ_τ^+ that we know from Proposition 6.1, the proof is complete.

$\square$

The next theorem describes the structure of an Erdős–Rényi graph conditional on the rare event that a nice graph parameter has a large deviation from its expected value. Following the convention introduced in Sect. 5.5 of Chap. 4, we will simply write G to denote the graphon f^G and the equivalence class $\tilde{f}^G$ of a simple graph G. Also, as in (5.5.2), we will use the notation $\delta_\square(G, \widetilde{F})$ to denote the infimum of $\delta_\square(G, \tilde{h})$ over all $\tilde{h} \in \widetilde{F}$.

Theorem 6.2 *Let τ and $G_{n,p}$ be as in Theorem 6.1. Take any $t \in \tau(\mathscr{W})$. Let $\widetilde{F}_\tau^+(p, t)$ be the set of minimizers of $I_p(\tilde{f})$ subject to the constraint $\tau(\tilde{f}) \geq t$ and let $\widetilde{F}_\tau^-(p, t)$ be the set of minimizers of $I_p(\tilde{f})$ subject to the constraint $\tau(\tilde{f}) \leq t$. Then $\widetilde{F}_\tau^+(p, t)$ and $\widetilde{F}_\tau^-(p, t)$ are nonempty compact subsets of $\widetilde{\mathscr{W}}$. Moreover, for any $\epsilon > 0$ there exist constants C^+ and C^- depending only on τ, p, t and ϵ such that*

$$\mathbb{P}(\delta_\square(G_{n,p}, \widetilde{F}_\tau^+(p, t)) \geq \epsilon \mid \tau(G_{n,p}) \geq t) \leq e^{-C^+ n^2},$$

$$\mathbb{P}(\delta_\square(G_{n,p}, \widetilde{F}_\tau^-(p, t)) \geq \epsilon \mid \tau(G_{n,p}) \leq t) \leq e^{-C^- n^2}.$$

The constant C^+ is positive if $\phi_\tau^+(p, t)$ is nonzero, and the constant C^- is positive if $\phi_\tau^-(p, t)$ is nonzero.

Proof This result is a simple consequence of Theorem 5.3 and Proposition 6.1. The condition (5.5.1) required for Theorem 5.3 can be easily shown to follow from the continuity of ϕ_τ^+ and ϕ_τ^- in t and the assumed positivity of these quantities, because any $\tilde{f}$ with $\tau(\tilde{f}) > t$ lies in the interior of the set $\{\tilde{h} : \tau(\tilde{h}) \geq t\}$. $\square$

6.4   Large Deviations for Subgraph Densities

In this section we will specialize the results of Sect. 6.3 to subgraph densities. Given two simple graphs G and H, recall the definition of the homomorphism density $t(H, G)$. Fix a simple graph H with at least one edge, and define

$$\phi^+(H, p, t) := \inf\{I_p(f) : f \in \mathscr{W}, \, t(H, f) \geq t\},$$

$$\phi^-(H, p, t) := \inf\{I_p(f) : f \in \mathscr{W}, \, t(H, f) \leq t\}.$$

Let $G_{n,p}$ be a random graph from the $G(n, p)$ model. The following theorem specializes Theorem 6.1 to subgraph densities.

Theorem 6.3 *Let H, $G_{n,p}$, ϕ^+ and ϕ^- be as above. Fix $p \in (0, 1)$. Then for any $t \in [0, 1]$,*

$$\lim_{n \to \infty} \frac{2}{n^2} \log \mathbb{P}(t(H, G_{n,p}) \geq t) = -\phi^+(H, p, t),$$

$$\lim_{n \to \infty} \frac{2}{n^2} \log \mathbb{P}(t(H, G_{n,p}) \leq t) = -\phi^-(H, p, t).$$

Proof Recall that by Exercise 3.1, for any graph G, $t(H, G)$ is the same as $t(H, f^G)$, where f^G is the graphon of G. The result now follows easily by Lemma 6.1 and Theorem 6.1. $\qquad\qquad\square$

Let $e(H)$ denote the number of edges in H. The following exercise describes the asymptotic behavior of $t(H, G_{n,p})$.

Exercise 6.3 Prove that $\mathbb{E}(t(H, G_{n,p})) = p^{e(H)}$, and that as $n \to \infty$, $t(H, G_{n,p})$ converges to $p^{e(H)}$ in probability.

The next lemma lists some basic properties of ϕ^+ and ϕ^-.

Lemma 6.3 *The functions ϕ^+ and ϕ^- and the related variational problems have the following properties:*

  (i) *The function ϕ^+ is continuous and non-decreasing in t, and the function ϕ^- is continuous and non-increasing in t.*
 (ii) *The minimum is attained in the variational problems defining ϕ^+ and ϕ^-.*
(iii) *For any t, if f minimizes $I_p(f)$ under the constraint $t(H, f) \geq t$, then $f \geq p$ almost everywhere. If f minimizes $I_p(f)$ under the constraint $t(H, f) \leq t$, then $f \leq p$ almost everywhere.*
 (iv) *The function ϕ^+ is zero when $t \leq p^{e(H)}$ and strictly increasing in t when $t > p^{e(H)}$. Similarly, ϕ^- is zero when $t \geq p^{e(H)}$ and strictly decreasing in t when $t < p^{e(H)}$.*

Proof The continuity and monotonicity claims follow from Proposition 6.1. The existence of minimizers follow from Theorem 6.2. The graphon $f \equiv p$ satisfies

$I_p(f) = 0$ and $t(H,f) = p^{e(H)}$. Since $I_p \geq 0$ everywhere, this shows that $\phi^+(H,p,t) = 0$ if $t \leq p^{e(H)}$ and $\phi^-(H,p,t) = 0$ if $t \geq p^{e(H)}$.

Take any $f \in \mathcal{W}$ that minimizes $I_p(f)$ subject to $t(H,f) \geq t$. Let $f_0 := \max\{f, p\}$. Then

$$t(H,f_0) \geq t(H,f) \geq t.$$

Since I_p is a strictly decreasing function in $[0,p]$, $I_p(f_0) > I_p(f)$ unless $f \geq p$ almost everywhere. Since f minimizes I_p subject to the constraint $t(H,f) \geq t$, this shows that $f \geq p$ almost everywhere.

Similarly, take any $f \in \mathcal{W}$ that minimizes $I_p(f)$ subject to $t(H,f) \leq t$. Let $f_0 := \min\{f, p\}$. Then

$$t(H,f_0) \leq t(H,f) \leq t.$$

Since I_p is a strictly increasing function in $[p,1]$, $I_p(f_0) < I_p(f)$ unless $f \leq p$ almost everywhere. Since f minimizes I_p subject to the constraint $t(H,f) \leq t$, this shows that $f \leq p$ almost everywhere.

Next, take $t > s \geq p^{e(H)}$. Take any $f \in \mathcal{W}$ that minimizes $I_p(f)$ subject to the constraint $t(H,f) \geq t$. Then $f \geq p$ almost everywhere, as shown above. Moreover, $f > p$ on a set of positive Lebesgue measure, since $t(H,f) \geq t > p^{e(H)}$. Since I_p is strictly increasing in $[p,1]$, this shows that if $g = (1-\epsilon)f + \epsilon p$ for some $\epsilon \in (0,1)$, then $I_p(g) < I_p(f)$. But if ϵ is small enough, the $t(H,g) \geq s$. This shows that $\phi^+(H,p,s) < \phi^+(H,p,t)$. The strict monotonicity of ϕ^- follows by a similar argument. $\square$

The following theorem specializes Theorem 6.2 to subgraph densities.

Theorem 6.4 *Let H and $G_{n,p}$ be as above. Take any $t \in [0,1]$. Let $\widetilde{F}^+(H,p,t)$ be the set of minimizers of $I_p(\tilde{f})$ subject to the constraint $t(H,\tilde{f}) \geq t$ and let $\widetilde{F}^-(H,p,t)$ be the set of minimizers of $I_p(\tilde{f})$ subject to the constraint $t(H,\tilde{f}) \leq t$. Then $\widetilde{F}^+(H,p,t)$ and $\widetilde{F}^-(H,p,t)$ are nonempty compact subsets of $\widetilde{\mathcal{W}}$. Moreover, for any $\epsilon > 0$ there exist positive constants C^+ and C^- depending only on H, p, t and ϵ such that if $t > p^{e(H)}$, then*

$$\mathbb{P}(\delta_\square(G_{n,p}, \widetilde{F}^+(H,p,t)) \geq \epsilon \mid t(H,G_{n,p}) \geq t) \leq e^{-C^+ n^2},$$

and if $t < p^{e(H)}$, then

$$\mathbb{P}(\delta_\square(G_{n,p}, \widetilde{F}^-(H,p,t)) \geq \epsilon \mid t(H,G_{n,p}) \leq t) \leq e^{-C^- n^2}.$$

Proof This theorem is an immediate corollary of Theorem 6.2. The positivity of C^+ and C^- follow from Theorem 6.2 and Lemma 6.3. $\square$

At this point, the obvious next step is to try to understand the nature of the sets $\widetilde{F}^+(H,p,t)$ and $\widetilde{F}^-(H,p,t)$ and to explicitly evaluate $\phi^+(H,p,t)$ and $\phi^-(H,p,t)$

if possible. Although explicit solutions to these problems are not known for all
possible H, p and t, we have a substantial amount of information. This is discussed
in the next four sections.

6.5 Euler–Lagrange Equations

In this section we will derive the Euler–Lagrange equations for the variational
problems defining the rate functions ϕ^+ and ϕ^- for subgraph densities. The main
technical challenge is that the domain for these variational problems is not an open
set, so one has to take care of boundary effects. The Euler–Lagrange equations will
be used later to explicitly solve the variational problems in certain regimes.

For a finite simple graph H, let $V(H)$ and $E(H)$ denote the sets of vertices and
edges of H. As before, let $e(H)$ be the number of edges of H. Given a symmetric
Borel measurable function $h : [0, 1]^2 \to \mathbb{R}$, for each $\{r, s\} \in E(H)$ and each pair of
points $x_r, x_s \in [0, 1]$, define

$$\Delta_{H,r,s}h(x_r, x_s) := \int_{[0,1]^{|V(H)|-2}} \prod_{\substack{\{r',s'\}\in E(H) \\ \{r',s'\}\neq\{r,s\}}} h(x_{r'}, x_{s'}) \prod_{\substack{v\in V(H) \\ v\neq r,s}} dx_v.$$

For $x, y \in [0, 1]$ define

$$\Delta_H h(x, y) := \sum_{\{r,s\}\in E(H)} \Delta_{H,r,s}h(x, y). \tag{6.5.1}$$

For example, when H is a triangle, then $V(H) = \{1, 2, 3\}$ and

$$\Delta_{H,1,2}h(x, y) = \Delta_{H,1,3}h(x, y) = \Delta_{H,2,3}h(x, y) = \int_0^1 h(x, z)h(y, z)\, dz$$

and therefore $\Delta_H h(x, y) = 3 \int_0^1 h(x, z)h(y, z)dz$. When H contains exactly one edge,
define $\Delta_H h \equiv 1$ for any h, by the usual convention that the empty product is 1.

Theorem 6.5 *Fix $p \in (0, 1)$, a finite simple graph H containing at least one edge,
and a number $t \in (0, 1)$. Let $\widetilde{F}^+ (H, p, t)$ and $\widetilde{F}^- (H, p, t)$ be defined as in the
statement of Theorem 6.4. Let $\alpha := \log(p/(1 - p))$ and Δ_H be defined as above.
Then for any $h \in \mathcal{W}$ such that $\tilde{h} \in \widetilde{F}^+ (H, p, t)$ or $\tilde{h} \in \widetilde{F}^- (H, p, t)$, there exists $\beta \in \mathbb{R}$
such that for almost every $(x, y) \in [0, 1]^2$,*

$$h(x, y) = \frac{e^{\alpha+\beta\Delta_H h(x,y)}}{1 + e^{\alpha+\beta\Delta_H h(x,y)}}.$$

Proof Fix h such that $\tilde{h} \in \widetilde{F}^+(H, p, t) \cup \widetilde{F}^-(H, p, t)$. The strict monotonicities of ϕ^+ and ϕ^- (Lemma 6.3) imply that $t(H, h) = t$. Therefore h minimizes $I_p(f)$ among all f satisfying $t(H, f) = t$. Let g be a bounded symmetric Borel measurable function from $[0, 1]$ into $\mathbb{R}$. For each $u \in \mathbb{R}$, let

$$f_u(x, y) := h(x, y) + u\, g(x, y),$$

and let $h_u := \alpha(u)f_u$, where

$$\alpha(u) = \left(\frac{t(H, h)}{t(H, f_u)} \right)^{1/e(H)}.$$

Note that $t(H, h_u) = t(H, h)$ for any u such that $t(H, f_u) \neq 0$. First suppose that h is bounded away from 0 and 1. Then $t(H, f_u) \neq 0$ and $h_u \in \mathcal{W}$ for every u sufficiently small in magnitude. Thus,

$$\frac{d}{du} I_p(h_u) \Big|_{u=0} = 0. \tag{6.5.2}$$

(Using the assumption that f is bounded away from 0 and 1, it is easy to check that $I_p(h_u)$ is differentiable in u for any h and g when $|u|$ is small enough.) A simple computation shows that

$$\frac{d}{du} I_p(h_u) \Big|_{u=0} = \int_{[0,1]^2} I_p'(h(x, y))(\alpha'(0)h(x, y) + \alpha(0)g(x, y))\, dx\, dy. \tag{6.5.3}$$

Note that

$$\frac{d}{du} t(H, f_u)$$

$$= \int_{[0,1]^{V(H)}} \sum_{\{r,s\} \in E(H)} g(x_r, x_s) \prod_{\substack{\{r',s'\} \in E(H) \\ \{r',s'\} \neq \{r,s\}}} f_u(x_{r'}, x_{s'}) \prod_{v \in V(H)} dx_v$$

$$= \int_{[0,1]^2} g(x, y) \Delta_H f_u(x, y)\, dy\, dx.$$

Now $\Delta_H f_u = \Delta_H h$ when $u = 0$. Thus, whenever g is such that

$$\int_{[0,1]^2} g(x, y) \Delta_H h(x, y)\, dy\, dx = 0, \tag{6.5.4}$$

then $\alpha'(0) = 0$, and hence by (6.5.2) and (6.5.3),

$$0 = \frac{d}{du} I_p(h_u) \Big|_{u=0} = \int_{[0,1]^2} I_p'(h(x, y))g(x, y)\, dx\, dy. \tag{6.5.5}$$

We claim that this implies that there exists $\beta \in \mathbb{R}$ such that for almost all $(x, y) \in [0, 1]^2$,

$$I'_p(h(x, y)) = \beta \Delta_H h(x, y), \tag{6.5.6}$$

which is the same as the assertion of the theorem. To prove this, first suppose that $\Delta_H h(x, y) = 0$ almost everywhere. Then taking $g(x, y) = I'_p(h(x, y))$, we see that (6.5.4) is satisfied, and hence (6.5.5) holds. This, in turn, implies that $I'_p(h(x, y)) = 0$ almost everywhere, and therefore (6.5.6) holds with any value of β.

Next, suppose that $\Delta_H h$ is nonzero on a set of positive measure. Define

$$\beta := \frac{\int_{[0,1]^2} I'_p(h(x, y)) \Delta_H h(x, y) \, dx \, dy}{\int_{[0,1]^2} \Delta_H h(x, y)^2 \, dx \, dy},$$

and

$$g(x, y) := I'_p(h(x, y)) - \beta \Delta_H h(x, y).$$

By the definition of β, (6.5.4) holds. Therefore (6.5.5) also holds. Subtracting β times the left-hand side of (6.5.4) from the left-hand side of (6.5.5) shows that (6.5.6) is satisfied almost everywhere.

Note that the above proof was carried out under the assumption that h is bounded away from 0 to 1. We will now prove that this assumption holds. For this proof, it is important to recall some basic properties of I_p, namely, that I_p is convex on $[0, 1]$, $I_p(p) = 0$, I_p is strictly increasing in $[p, 1]$ and strictly decreasing in $[0, p]$, $I'_p(x) \to -\infty$ as $x \to 0$, and $I'_p(x) \to \infty$ as $x \to 1$.

First, suppose that $\tilde{h} \in \widetilde{F}^-(H, p, t)$. By Lemma 6.3, $h \leq p$ almost everywhere. So we only have to show that h is bounded away from zero. Fix $\epsilon > 0$ and let

$$w(x, y) := \begin{cases} 0 & \text{if } h(x, y) \geq \epsilon, \\ 1 & \text{if } h(x, y) < \epsilon. \end{cases}$$

Suppose that $h < \epsilon$ on a set of positive Lebesgue measure, so that

$$\int_{[0,1]^2} w(x, y) \, dx \, dy > 0. \tag{6.5.7}$$

For each $u \geq 0$, let

$$g_u(x, y) := (1 - Au)(h(x, y) + u w(x, y)),$$

where

$$A := B \int_{[0,1]^2} w(x, y) \, dx \, dy,$$

where B is a constant, to be chosen later. We have already observed at the beginning of this proof that $t(H, h) = t$ by a consequence of Lemma 6.3. A simple computation using this fact shows that

$$\frac{d}{du}t(H, g_u)\Big|_{u=0} = -e(H)At(H, h) + \int_{[0,1]^2} w(x, y)\Delta_H h(x, y)\,dx\,dy$$

$$= \int_{[0,1]^2} (-e(H)Bt + \Delta_H h(x, y))w(x, y)\,dx\,dy.$$

Choose B so large (depending only on t, H and h) such that for all x, y,

$$-e(H)Bt + \Delta_H h(x, y) < 0.$$

Then by (6.5.7) and the above identity,

$$\frac{d}{du}t(H, g_u)\Big|_{u=0} < 0. \tag{6.5.8}$$

Now note that

$$\frac{d}{du}I_p(g_u)\Big|_{u=0} = \int_{[0,1]^2} I_p'(h(x, y))(-Ah(x, y) + w(x, y))\,dx\,dy \tag{6.5.9}$$

$$\leq \int_{[0,1]^2} (-CB + I_p'(h(x, y)))w(x, y)\,dx\,dy,$$

where

$$C = \int_{[0,1]^2} I_p'(h(x, y))h(x, y)\,dx\,dy.$$

It is easy to see that C is finite, using the fact that $h \leq p$ almost everywhere. Note also that C does not depend on ϵ. Therefore, if ϵ is so small that

$$I_p'(\epsilon) < CB, \tag{6.5.10}$$

then by (6.5.9) and (6.5.7) (and the properties of I_p listed before),

$$\frac{d}{du}I_p(g_u)\Big|_{u=0} < 0. \tag{6.5.11}$$

Since h is bounded away from 1, $g_u \in \mathcal{W}$ for all sufficiently small positive u. Therefore by the minimizing property of h, the inequalities (6.5.8) and (6.5.11) cannot hold simultaneously. Therefore, if ϵ is so small that (6.5.10) is satisfied, then (6.5.7) must be invalid. In other words, $h \geq \epsilon$ almost everywhere. This completes the proof that any $\tilde{h} \in \widetilde{F}^-(H, p, t)$ is bounded away from 0 and 1.

Next, take any h such that $\tilde{h} \in \widetilde{F}^+(H, p, t)$. The proof is quite similar to the previous case, with minor modifications. By Lemma 6.3, $h \geq p$ almost everywhere. So we only have to show that h is bounded away from 1. Fix $\epsilon > 0$ and let

$$
w(x, y) := \begin{cases} 0 & \text{if } h(x, y) \leq 1 - \epsilon, \\ 1 & \text{if } h(x, y) > 1 - \epsilon. \end{cases}
$$

Suppose that $h > 1 - \epsilon$ on a set of positive Lebesgue measure, so that

$$
\int_{[0,1]^2} w(x, y) \, dx \, dy > 0. \tag{6.5.12}
$$

For each $u \geq 0$, let

$$
g_u(x, y) := (1 - Au)(h(x, y) - u \, w(x, y)) + Au,
$$

where

$$
A := B \int_{[0,1]^2} w(x, y) \, dx \, dy,
$$

where B is a constant, to be chosen later. A simple computation shows that

$$
\frac{d}{du} t(H, g_u) \Big|_{u=0} = AD - \int_{[0,1]^2} w(x, y) \Delta_H h(x, y) \, dx \, dy \tag{6.5.13}
$$

$$
= \int_{[0,1]^2} (BD - \Delta_H h(x, y)) w(x, y) \, dx \, dy,
$$

where

$$
D = \sum_{\{r,s\} \in E(H)} \int_{[0,1]^{V(H)}} (1 - h(x_r, x_s)) \prod_{\substack{\{r',s'\} \in E(H) \\ \{r',s'\} \neq \{r,s\}}} h(x_{r'}, x_{s'}) \prod_{v \in V(H)} dx_v.
$$

Since $h \geq p$ almost everywhere, the above formula shows that D can be 0 only if $h = 1$ almost everywhere. Since $t(H, h) = t < 1$, this is not true. Therefore $D > 0$. Thus, B can be chosen (depending only on t, H and h) such that for all x, y,

$$
BD - \Delta_H h(x, y) > 0.
$$

Then by (6.5.12) and (6.5.13),

$$\frac{d}{du} t(H, g_u)\Big|_{u=0} > 0. \tag{6.5.14}$$

Now note that

$$\frac{d}{du} I_p(g_u)\Big|_{u=0} = \int_{[0,1]^2} I_p'(h(x,y))(A(1 - h(x,y)) - w(x,y))\, dx\, dy \tag{6.5.15}$$

$$\leq \int_{[0,1]^2} (CB - I_p'(h(x,y)))w(x,y)\, dx\, dy,$$

where

$$C = \int_{[0,1]^2} I_p'(h(x,y))(1 - h(x,y))\, dx\, dy.$$

It is easy to see that C is finite, using the fact that $h \geq p$ almost everywhere. Note also that C does not depend on ϵ. Therefore, if ϵ is so small that

$$I_p'(1 - \epsilon) > CB, \tag{6.5.16}$$

then by (6.5.15) and (6.5.12) (and the properties of I_p listed before),

$$\frac{d}{du} I_p(g_u)\Big|_{u=0} < 0. \tag{6.5.17}$$

Since h is bounded away from 0, $g_u \in \mathscr{W}$ for all sufficiently small positive u. Therefore by the minimizing property of h, the inequalities (6.5.14) and (6.5.17) cannot hold simultaneously. Therefore, if ϵ is so small that (6.5.16) is satisfied, then (6.5.12) must be invalid. In other words, $h \leq 1 - \epsilon$ almost everywhere. This completes the proof that any $\tilde{h} \in \widetilde{F}^+(H, p, t)$ is bounded away from 0 and 1.   □

6.6   The Symmetric Phase

Let ϕ^+ and ϕ^- be the rate functions for subgraph densities, defined in Sect. 6.4. Fix H and p. Take any $t > p^{e(H)}$. We will say that t is in the symmetric phase if the set of minimizers $\widetilde{F}^+(H, p, t)$ consists of a unique constant function. Similarly, for $t < p^{e(H)}$, we will say that t is in the symmetric phase if the set of minimizers $\widetilde{F}^-(H, p, t)$ consists of a unique constant function. The following result clarifies the significance of the symmetric phase.

Proposition 6.2 *Let H be a finite simple graph with at least one edge and p be an element of $(0, 1)$. Suppose that $t > p^{e(H)}$ belongs to the symmetric phase. Let $r := t^{1/e(H)}$. Let $G_{n,p}$ and $G_{n,r}$ be independent Erdős–Rényi graphs defined on the same probability space. Then for any $\epsilon > 0$,*

$$\lim_{n \to \infty} \mathbb{P}(\delta_\square(G_{n,p}, G_{n,r}) > \epsilon \mid t(H, G_{n,p}) \geq t) = 0.$$

Similarly, if $t < p^{e(H)}$ belongs to the symmetric phase, then

$$\lim_{n \to \infty} \mathbb{P}(\delta_\square(G_{n,p}, G_{n,r}) > \epsilon \mid t(H, G_{n,p}) \leq t) = 0.$$

Proof This result is an easy consequence of the definition of symmetric phase, Theorem 6.2 and Exercise 5.1. □

The next theorem, which is the main result of this section, says that if t is close enough to $p^{e(H)}$, it is in the symmetric phase. The proof uses the Euler–Lagrange equations derived in the previous section.

Theorem 6.6 *Take any finite simple graph H consisting of at least one edge and a number $p \in (0, 1)$. For any $t \in [0, 1]$, let $c_{H,t}$ be the graphon that is identically equal to $t^{1/e(H)}$. Then there exists $\delta > 0$ depending only on H and p such that $\widetilde{F}^+(H, p, t) = \{\tilde{c}_{H,t}\}$ if $p^{e(H)} < t < p^{e(H)} + \delta$ and $\widetilde{F}^-(H, p, t) = \{\tilde{c}_{H,t}\}$ if $p^{e(H)} - \delta < t < p^{e(H)}$. Consequently, in the first case, $\phi^+(H, p, t) = I_p(t^{1/e(H)})$ and in the second case, $\phi^-(H, p, t) = I_p(t^{1/e(H)})$.*

Proof In this proof, C will throughout denote any function from $[0, \infty)$ into $[0, \infty]$, depending only on p and H but not on t, such that

$$\lim_{x \to 0} C(x) = 0.$$

The function C may change from line to line. Let

$$\epsilon := |t - p^{e(H)}|.$$

We claim that for any h such that $\tilde{h} \in \widetilde{F}^+(H, p, t) \cup \widetilde{F}^-(H, p, t)$,

$$\int_{[0,1]^2} (h(x, y) - p)^2 \, dx \, dy < C(\epsilon). \tag{6.6.1}$$

To see this, first observe that $t(H, c_{H,t}) = t$ and $I_p(c_{H,t}) < C(\epsilon)$. Therefore, for any h such that $\tilde{h} \in \widetilde{F}^+(H, p, t) \cup \widetilde{F}^-(H, p, t)$,

$$I_p(h) < C(\epsilon). \tag{6.6.2}$$

The map I_p on $[0, 1]$ satisfies $I_p(p) = I_p'(p) = 0$ and

$$I_p''(x) = \frac{1}{2x(1-x)} \geq 2 \text{ for all } x \in [0, 1],$$

and therefore

$$I_p(x) \geq (x-p)^2 \text{ for all } x \in [0, 1].$$

This, combined with (6.6.2), proves (6.6.1). Now fix any h such that

$$\tilde{h} \in \widetilde{F}^+(H, p, t) \cup \widetilde{F}^-(H, p, t).$$

By Theorem 6.5, there is some $\beta \in \mathbb{R}$ such that h satisfies

$$\log \frac{h(x, y)}{1 - h(x, y)} - \log \frac{p}{1-p} = \beta \Delta_H h(x, y) \tag{6.6.3}$$

for almost all x, y.

It is easy to see from (6.6.1) that

$$\int_{[0,1]^2} (\Delta_H h(x, y) - p^{e(H)-1})^2 \, dx \, dy < C(\epsilon).$$

From this and (6.6.1), it follows that there exists $(x, y) \in [0, 1]^2$ that satisfies (6.6.3), such that

$$\left| \log \frac{h(x, y)}{1 - h(x, y)} - \log \frac{p}{1-p} \right| < C(\epsilon)$$

and

$$|\Delta_H h(x, y) - p^{e(H)-1}| < C(\epsilon).$$

Consequently, from (6.6.3), we get

$$|\beta| < C(\epsilon). \tag{6.6.4}$$

Let $\| \cdot \|_\infty$ denote the L^∞ norm on $\mathscr{W}$. Let σ be any measure preserving bijection of $[0, 1]$ and let $g(x, y) := h(\sigma x, \sigma y)$. Then g also satisfies (6.6.3). A simple computation shows that

$$\|\Delta_H h - \Delta_H g\|_\infty \leq \sum_{\{r,s\} \in E(H)} \|\Delta_{H,r,s} h - \Delta_{H,r,s} g\|_\infty$$

$$\leq e(H)(e(H) - 1)\|h - g\|_\infty.$$

Let $\alpha := \log(p/(1-p))$. Using the above inequality, Theorem 6.5 and the inequality

$$\left| \frac{e^a}{1+e^a} - \frac{e^b}{1+e^b} \right| \leq \frac{|a-b|}{4}$$

(easily proved by the mean value theorem) it follows that for almost all x, y,

$$\begin{aligned}
|h(x, y) - g(x, y)| &= \left| \frac{e^{\alpha + \beta \Delta_H h(x,y)}}{1 + e^{\alpha + \beta \Delta_H h(x,y)}} - \frac{e^{\alpha + \beta \Delta_H g(x,y)}}{1 + e^{\alpha + \beta \Delta_H g(x,y)}} \right| \\
&\leq \frac{1}{4} |\beta| \| \Delta_H h - \Delta_H g \|_\infty \\
&\leq \frac{1}{4} \| h - g \|_\infty |\beta| e(H)(e(H) - 1).
\end{aligned}$$

If the coefficient of $\| h - g \|_\infty$ in the last expression is strictly less than 1, it follows that h must be equal to g almost everywhere. Since this would hold for any bijection σ, h must be a constant function. Combined with (6.6.4), this completes the proof.
□

6.7   Symmetry Breaking

Let all notation be as in Sect. 6.4. Let H be a finite simple graph. Take any $p \in (0, 1)$. We will say that a number $t > p^{e(H)}$ belongs to the region of broken symmetry if $\widetilde{F}^+(H, p, t)$ contains only non-constant graphons, and a number $t < p^{e(H)}$ belongs to the region of broken symmetry if $\widetilde{F}^-(H, p, t)$ contains only non-constant graphons.

Note that the definition leaves open the possibility that a number t may belong to neither the symmetric phase nor the region of broken symmetry, for example if $\widetilde{F}^+(H, p, t)$ contains both constant and non-constant graphons. The following theorem shows that a region of broken symmetry for the upper tail exists under a mild condition on H. Recall that the degree of a vertex in a graph is the number of vertices that are adjacent to it. The average degree of a graph is the average of the vertex degrees.

Theorem 6.7 *Let H be a finite simple graph with average degree strictly greater than one. Let $\widetilde{C}$ denote the set of constant functions in $\widetilde{\mathscr{W}}$. Then for each $t \in (0, 1)$, there exists $p' > 0$ such that for all $0 < p < p'$, $\widetilde{F}^+(H, p, t) \cap \widetilde{C} = \emptyset$. Moreover, for such p, there exists $\epsilon > 0$ such that*

$$\lim_{n \to \infty} \mathbb{P}(\delta_\square(G_{n,p}, \widetilde{C}) > \epsilon \mid t(H, G_{n,p}) \geq t) = 1.$$

Proof Take any $t \in (0, 1)$. Let $v(H)$ be the number of vertices of H, and let

$$\chi_{H,t}(x, y) := \begin{cases} 1 & \text{if } \max\{x, y\} \leq t^{1/v(H)}, \\ 0 & \text{otherwise.} \end{cases}$$

Then $t(H, \chi_{H,t}) \geq t$, and for any $t \in (0, 1)$,

$$\lim_{p \to 0} \frac{I_p(c_{H,t})}{\log(1/p)} = t^{1/e(H)} > t^{2/v(H)} = \lim_{p \to 0} \frac{I_p(\chi_{H,t})}{\log(1/p)},$$

by the assumed condition that the average degree is strictly greater than one (which is the same as saying $2e(H)/v(H) > 1$). Moreover, $t(H, c_{H,t'}) \geq t$ if and only if $t' \geq t$, and if $t' \geq t$, then $I_p(c_{H,t'}) \geq I_p(c_{H,t})$. Thus, the above inequality shows that for each $t \in (0, 1)$, there exists $p' > 0$ such that for all $0 < p < p'$ we have

$$\widetilde{F}^+(H, p, t) \cap \widetilde{C} = \emptyset.$$

Since $\widetilde{F}^+(H, p, t)$ and $\widetilde{C}$ are non-empty compact subsets of $\widetilde{\mathcal{W}}$, this implies that

$$\delta_\square(\widetilde{F}^+(H, p, t), \widetilde{C}) > 0.$$

It is now easy to complete the proof using Theorem 6.2. $\qquad \square$

If H is a simple graph and H' is obtained from H by eliminating isolated vertices (if any), then $t(H', G) = t(H, G)$ for any G. Therefore in our study of subgraph densities we may consider only those H that have no isolated vertices. If H is a finite simple graph with no isolated vertices, then it is easy to see that the average degree of H is at least one. The following exercise shows that for the existence of region of broken symmetry in the upper tail, it is necessary that the average degree is strictly bigger than one.

Exercise 6.4 If H has no isolated vertices and has average degree exactly equal to one, show that for any $p \in (0, 1)$, there is no region of broken symmetry in the upper tail.

Notice that Theorem 6.7 only says that regions of broken symmetry exist for small enough p. We will see later that unless p is small enough (depending on H), a region of broken symmetry in the upper tail may not exist.

The next theorem is the analog of Theorem 6.7 for lower tails. Recall that the chromatic number of a graph is the minimum number of colors required to color the vertices such that no two vertices connected by an edge receive the same color.

Theorem 6.8 *Let H be a finite simple graph with chromatic number at least three. Then for each $p \in (0, 1)$, there exists $t' \in (0, 1)$ such that for all $0 < t < t'$, $\widetilde{F}^-(H, p, t) \cap \widetilde{C} = \emptyset$. Moreover, for such t, there exists $\epsilon > 0$ such that*

$$\lim_{n \to \infty} \mathbb{P}(\delta_\square(G_{n,p}, \widetilde{C}) > \epsilon \mid t(H, G_{n,p}) \leq t) = 1.$$

Proof Let k be a chromatic number of H. For a real number x, let $\lfloor x \rfloor$ denote the largest integer that is $\leq x$. Define

$$\psi_{H,p}(x, y) := \begin{cases} p & \text{if } \lfloor (k-1)x \rfloor \neq \lfloor (k-1)y \rfloor, \\ 0 & \text{otherwise.} \end{cases}$$

Let n be the number of vertices of H. Label these vertices as $1, 2, \ldots n$. Suppose that $x_1, \ldots, x_n \in (0, 1)$ are points such that $\psi_{H,p}(x_i, x_j) \neq 0$ for all $\{i, j\} \in E(H)$. Let $r_i : \lfloor (k - 1)x_i \rfloor$. Then by the definition of $\psi_{H,p}$, $r_i \neq r_j$ for all $\{i, j\} \in E(H)$. Thus, if we color vertex i with color r_i, then no two adjacent vertices receive the same color. However, r_i can take only $k - 1$ possible values. This contradicts the fact that k is the chromatic number of H. Therefore, such $x_1, \ldots, x_n$ cannot exist. This proves that

$$t(H, \psi_{H,p}) = 0. \tag{6.7.1}$$

Since $k \geq 3$ and $I_p(p) = 0$,

$$\lim_{t \to 0} \frac{I_p(c_{H,t})}{\log(1/(1 - p))} = 1 > \frac{1}{k - 1} = \frac{I_p(\psi_{H,p})}{\log(1/(1 - p))}.$$

By (6.7.1), by a similar argument as in the proof of Theorem 6.7, this shows that for all sufficiently small t,

$$\widetilde{F}^-(H, p, t) \cap \widetilde{C} = \emptyset.$$

The rest of the proof is similar to that of Theorem 6.7.   □

A graph with chromatic number one is just a collection of isolated vertices, and therefore quite uninteresting. It is conjectured that a graph with chromatic number two cannot have a region of broken symmetry in the lower tail, which means that the condition on H in Theorem 6.8 is necessary for the existence of a region of broken symmetry. The conjecture has been proved in some special cases. We will discuss more about this later.

6.8 The Phase Boundary for Regular Subgraphs

Let us continue to work in the setting of Sect. 6.4. The additional assumption in this section is that the graph H is regular. (Recall that a graph is called regular if all its vertices have the same degree.) When H is regular, the exact boundary between the symmetric phase and the phase of broken symmetry for the upper tail can be identified. This is the content of the following theorem. To understand the statement of the theorem, recall that the convex minorant of a function f is the greatest convex function g such that $g \leq f$ everywhere on the domain of f. Note that this definition makes sense because the pointwise supremum of an arbitrary collection of convex functions is convex. Recall also the definitions of symmetric phase and symmetry breaking from Sects. 6.6 and 6.7.

Theorem 6.9 *Let H be a finite, simple, regular graph of degree $d \geq 2$. Take any $p \in (0, 1)$ and $t \in (p^{e(H)}, 1)$. Let $r := t^{1/e(H)}$. Then t belongs to the symmetric phase if the point $(r^d, I_p(r))$ lies on the convex minorant of the function $J_p(x) := I_p(x^{1/d})$.*

On the other hand, if $(r^d, I_p(r))$ does not lie on the convex minorant of J_p, then t belongs to the region of broken symmetry.

To prove this theorem, we need a small amount of preparation.

Lemma 6.4 *Let J_p'' denote the second derivative of J_p. Then J_p'' cannot have more than two zeroes in $(0, 1)$.*

Proof A simple computation gives

$$I_p'(x) = \log \frac{x}{1 - x} - \log \frac{p}{1 - p}$$

and

$$I_p''(x) = \frac{1}{x(1 - x)}.$$

Thus,

$$J_p''(x) = \frac{1}{d}\left(\frac{1}{d} - 1\right)x^{1/d-2}I_p'(x^{1/d}) + \frac{1}{d}x^{2/d-2}I_p''(x^{1/d}),$$

which implies that $J_p''(x) = 0$ if and only if

$$(d - 1)I_p'(x^{1/d}) = x^{1/d}I_p''(x^{1/d}).$$

In other words, x is a zero of J_p'' if and only if $x^{1/d}$ is a zero of

$$K_p(y) := (d - 1)I_p'(y) - yI_p''(y).$$

Now note that

$$K_p'(y) = \frac{d - 1}{y(1 - y)} - \frac{1}{(1 - y)^2} = \frac{(d - 1)(1 - y) - y}{y(1 - y)^2},$$

which shows that K_p' has exactly one zero in $(0, 1)$. By Rolle's theorem, this implies that K_p can have at most two zeroes in $(0, 1)$. This completes the proof of the lemma. $\square$

Lemma 6.5 *Take any $r \in (p, 1)$. If the point $(r^d, I_p(r))$ lies on the graph of the convex minorant $\hat{J}_p$ of J_p, then $\hat{J}_p$ cannot be linear in a neighborhood of r^d.*

Proof Suppose that $\hat{J}_p$ is linear in a neighborhood of r^d. First, note that by the formulas for I_p', I_p'' and J_p'' derived in the proof of Lemma 6.4, it follows that $J_p''(x) > 0$ when $x = p^d$ and also when x is sufficiently close to 1.

Next, recall that $J_p(r^d) = \hat{J}_p(r^d)$, $\hat{J}_p$ is linear in a neighborhood of r^d and $J_p \geq \hat{J}_p$ everywhere. From these it follows easily that $J_p'(r^d) = \hat{J}_p'(r^d)$ and $J_p''(r^d) \geq \hat{J}_p''(r^d) = 0$.

Now suppose that J_p is convex in (p^d, r^d). Then by the given conditions, the function that equals J_p in $(0, r^d]$ and equals $\hat{J}_p$ in $[r^d, 1)$ is a convex function (since its derivative is nondecreasing) and lies between $\hat{J}_p$ and J_p everywhere. Therefore it must be equal to $\hat{J}_p$ everywhere. Thus, the convexity of J_p in (p^d, r^d) would imply that $J_p = \hat{J}_p$ in (p^d, r^d). But by Lemma 6.4, J_p cannot be linear in an interval. Therefore the above scenario is impossible; J_p cannot be convex in (p^d, r^d). In particular, there exists a point in this interval where J'' is strictly negative.

By a similar argument, J_p cannot be convex in $(r^d, 1)$, and therefore there exists a point in this interval where J_p'' is strictly negative.

Let us now collect our deductions. Under the assumption that $\hat{J}_p$ is linear in a neighborhood of r^d, we have argued that $J_p''(r^d) \geq 0$, and there exist $x_1 \in (p^d, r^d)$ and $x_2 \in (r^d, 1)$ such that $J_p''(x_1) < 0$ and $J_p''(x_2) < 0$. We have also observed that $J_p''(x) > 0$ for x sufficiently close to p^d and for x sufficiently close to 1. These deductions jointly imply that J_p'' must have at least three zeroes, which is impossible by Lemma 6.4. $\qquad\square$

Lemma 6.6 *If $(r^d, I_p(r))$ does not lie on the convex minorant of J_p, then there exist $0 < r_1 < r < r_2 < 1$ such that $(r^d, I_p(r))$ lies strictly above the line segment joining $(r_1^d, I_p(r_1))$ and $(r_2^d, I_p(r_2))$.*

Proof First, note that the function that is identically equal to zero on $(0, 1)$ is a convex function that lies below J_p, and equals J_p at p^d. Thus, $J_p(p^d) = \hat{J}_p(p^d)$. Next, note that by the formula for J_p'' from the proof of Lemma 6.4, we know that J_p is convex near 1. Moreover, it is easy to verify that $J_p'(x) \to \infty$ as $x \to 1$. From these two facts and the observation that $\hat{J}_p$ is bounded below on $(0, 1)$, it follows easily that if x is sufficiently close to 1, then the tangent line to J_p at x lies entirely below J_p in the interval $(0, 1)$. Consequently, for such x, $J_p(x) = \hat{J}_p(x)$. To summarize, $J_p = \hat{J}_p$ at p^d and at all x sufficiently close to 1.

Let r_1 be the largest number less than r such that $J_p(r_1^d) = \hat{J}_p(r_1^d)$ and r_2 be the smallest number bigger than r such that $J_p(r_2^d) = \hat{J}_p(r_2^d)$. Choose $a \in (r_1, r)$ and $b \in (r, r_2)$. Let h be the continuous function that equals $\hat{J}_p$ in $(0, a^d) \cup (b^d, 1)$, and is linear in the interval $[a^d, b^d]$. Since $\hat{J}_p$ is convex, so is h. Moreover, $h \geq \hat{J}_p$ everywhere. For each $s \in [0, 1]$, let $h_s := sh + (1 - s)\hat{J}_p$. Then for each s, h_s is a convex function lying between $\hat{J}_p$ and h. Since $J_p > \hat{J}_p$ in the compact interval $[a^d, b^d]$, there must exist $\epsilon > 0$ such that $J_p(x) \geq \hat{J}_p(x) + \epsilon$ for all x in this interval. Thus, for small enough positive s, h_s lies below J_p. Since $\hat{J}_p$ is the convex minorant of J_p, this is possible only if $h = \hat{J}_p$ in $[a^d, b^d]$. In other words, $\hat{J}_p$ is linear in this interval. Taking $a \to r_1$ and $b \to r_2$, we see that $\hat{J}_p$ is linear in (r_1, r_2). This completes the proof of the lemma. $\qquad\square$

We are now ready to prove Theorem 6.9.

Proof (Proof of Theorem 6.9) First, suppose that $(r^d, I_p(r))$ lies on the convex minorant of J_p, which we denote by $\hat{J}_p$. Recall the generalized Hölder's inequality (Theorem 2.1) proved in Sect. 2.5 of Chap. 2. By that inequality and the regularity of H, it follows that for any $f \in \mathcal{W}$,

$$t(H,f) \le \left(\int_{[0,1]^2} f(x,y)^d \, dx \, dy \right)^{e(H)/d}.$$

If $t(H,f) \ge t = r^{e(H)}$, then by the above inequality, Jensen's inequality and the assumption that $\hat{J}_p(r^d) = J_p(r^d)$, we get

$$I_p(f) = \int_{[0,1]^2} J_p(f(x,y)^d) \, dx \, dy \ge \int_{[0,1]^2} \hat{J}_p(f(x,y)^d) \, dx \, dy \tag{6.8.1}$$

$$\ge \hat{J}_p \left(\int_{[0,1]^2} f(x,y)^d \, dx \, dy \right) \ge \hat{J}_p(r^d) = J_p(r^d) = I_p(r).$$

Moreover, equality holds if and only if $f = r$ almost everywhere since $\hat{J}_p$ is not linear in any neighborhood of r^d by Lemma 6.5 and $\hat{J}_p$ is strictly increasing in $(p^d, 1)$ (which is easy to prove using the properties of J_p). This proves the first part of the theorem.

Next, suppose that $(r^d, I_p(r))$ does not lie on the convex minorant of J_p. Then by Lemma 6.6, there exists $0 < r_1 < r < r_2 < 1$ and $s \in (0, 1)$ such that

$$sr_1^d + (1 - s)r_2^d = r^d$$

and

$$sI_p(r_1) + (1 - s)I_p(r_2) < I_p(r). \tag{6.8.2}$$

Choose some $\epsilon > 0$ and define

$$a := s\epsilon^2, \quad b := (1 - s)\epsilon^2 + \epsilon^3.$$

Let ϵ be chosen so small that $0 < a < 1 - b < 1$. Define three intervals $I_0 := (a, 1 - b)$, $I_1 := (0, a)$ and $I_2 : (1 - b, 1)$. Define a graphon

$$f_\epsilon(x,y) := \begin{cases} r_1 & \text{if } (x,y) \in (I_0 \times I_1) \cup (I_1 \times I_0), \\ r_2 & \text{if } (x,y) \in (I_0 \times I_2) \cup (I_2 \times I_0), \\ r & \text{in all other cases.} \end{cases}$$

Then note that

$$I_p(f_\epsilon) = 2a(1 - a - b)(I_p(r_1) - I_p(r)) + 2b(1 - a - b)(I_p(r_2) - I_p(r))$$
$$= 2(1 - a - b)(aI_p(r_1) + bI_p(r_2) - (a + b)I_p(r))$$
$$= 2(1 - a - b)\epsilon^2(sI_p(r_1) + (1 - s)I_p(r_2) - I_p(r) + (I_p(r_2) - I_p(r))\epsilon).$$

By (6.8.2), this shows that $I_p(f_\epsilon) < I_p(r)$ if ϵ is chosen sufficiently small.

Let the vertices of H be labeled as $1, 2, \ldots, k$. Let $U_1, \ldots, U_k$ be independent random variables that are uniformly distributed in the interval $[0, 1]$. Then

$$t(H, f_\epsilon) = \mathbb{E}(S),$$

where

$$S := \prod_{\{i,j\} \in E(H)} f_\epsilon(U_i, U_j).$$

For each $1 \le i \le k$, let A_i be the event that $U_i \in I_1 \cup I_2$ and $U_j \notin I_1 \cup I_2$ for all $j \ne i$. Let B be the event that at least two of the U_i's belong to $I_1 \cup I_2$, and E be the event that none of the U_i's belong to $I_1 \cup I_2$. Since the length of $I_1 \cup I_2$ decays like ϵ^2 as $\epsilon \to 0$, the event B has probability $O(\epsilon^4)$ as $\epsilon \to 0$. On the event E, $S = r^{e(H)}$. On the event A_i,

$$f(U_i, U_j) = \begin{cases} r_1 & \text{if } U_i \in I_1 \text{ and } \{i,j\} \in E(H), \\ r_2 & \text{if } U_i \in I_2 \text{ and } \{i,j\} \in E(H), \\ r & \text{if } \{i,j\} \notin E(H). \end{cases}$$

Since H is a regular graph of degree d, this implies that as $\epsilon \to 0$,

$$\mathbb{E}(S; A_i) = (ar_1^d + br_2^d)r^{e(H)-d} + O(\epsilon^4),$$

where $\mathbb{E}(S; A_i)$ is the expectation of S times the indicator of the event A_i, as per usual convention. Combining the above observations, we get that as $\epsilon \to 0$,

$$t(H, f_\epsilon) - r^{e(H)} = k(a(r_1^d - r^d) + b(r_2^d - r^d))r^{e(H)-d} + O(\epsilon^4).$$

A simple calculations shows that

$$a(r_1^d - r^d) + b(r_2^d - r^d) = \epsilon^3(r_2^d - r^d).$$

Thus, $t(H, f_\epsilon) > t$ when ϵ is sufficiently small. Together with our previous deduction that $I_p(f_\epsilon) < I_p(r)$ for ϵ sufficiently small, this shows that t belongs to the region of broken symmetry.   □

6.9 The Double Phase Transition

Theorem 6.9 uncovers a remarkable property of Erdős–Rényi random graphs conditioned to have a large homomorphism density of some regular graph H. The property may be roughly described as follows. If the density of H is slightly more than the expected density, or if the density is close to 1, then the graph behaves just like an Erdős–Rényi graph with a different edge probability. If the density lies between these two thresholds, the behavior does not resemble that of an Erdős–Rényi graph. In the language of statistical physics, the system goes from a symmetric phase to a phase of broken symmetry as the density parameter increases, but then reverts back to the symmetric phase at a high enough value of the density parameter. This is a highly unusual behavior for a statistical mechanical system. The following theorem gives the precise statement.

Theorem 6.10 *Let H be a finite, simple, regular graph of degree $d \geq 2$. Take any $p \in (0, 1)$. Then there exists $p < r_1 \leq r_2 < 1$ such that the symmetric phase of the upper tail (as defined in Sect. 6.6) is the set $[p^d, r_1^d] \cup [r_2^d, 1]$, and the region of broken symmetry (as defined in Sect. 6.7) is the interval (r_1^d, r_2^d). If p is sufficiently small, then $r_1 < r_2$, which means that the region of broken symmetry is nonempty.*

Proof Let J_p be defined as in Theorem 6.9. First, suppose that J_p is coincides with $\hat{J}_p$ everywhere in $[p^d, 1]$. Then by Theorem 6.9, the region of broken symmetry is empty and the claimed result is valid.

Next, suppose that there is at least one $r \in (p, 1)$ such that $(r^d, I_p(r))$ does not lie on the convex minorant of J_p. In other words, $J_p(r^d) > \hat{J}_p(r^d)$. As in the proof of Lemma 6.6, let r_1 be the largest number less than r such that $J_p(r_1^d) = \hat{J}_p(r_1^d)$ and r_2 be the smallest number bigger than r such that $J_p(r_2^d) = \hat{J}_p(r_2^d)$. We saw in the proof of Lemma 6.6 that $\hat{J}_p$ is linear in the interval $[r_1^d, r_2^d]$. Since J_p and $\hat{J}_p$ coincide at r_1^d and r_2^d, it is easy to see that J_p'' is nonnegative at these two points. Since $\hat{J}_p$ is linear in (r_1^d, r_2^d) and J_p lies strictly above $\hat{J}_p$ in this interval, J_p'' must be strictly negative somewhere in this interval. Thus, J_p'' has at least two zeroes in $[r_1^d, r_2^d]$. If J_p and $\hat{J}_p$ do not coincide somewhere outside (r_1^d, r_2^d), a repetition of the above argument would imply the existence of more zeroes, going against the assertion of Lemma 6.4. Thus, $[p^d, r_1^d] \cup [r_2^d, 1]$ is the symmetric phase. Lastly, recall that by Theorem 6.7, the region of broken symmetry is nonempty if p is sufficiently small. $\qquad\square$

6.10 The Lower Tail and Sidorenko's Conjecture

Let us continue using the notation of Sect. 6.4. The object of interest now is the set $\widetilde{F}^-(H, p, t)$. We have already seen in Theorem 6.6 that if t is sufficiently close to $p^{e(H)}$, then $\widetilde{F}^-(H, p, t)$ consists of a single constant graphon. Theorem 6.8 tells us that if H has chromatic number at least three, then any sufficiently small t must belong to the region of broken symmetry. What about graphs of chromatic number two, that is, bipartite graphs? There is a famous conjecture about bipartite graphs, known as Sidorenko's conjecture, which claims that for any bipartite graph H and any finite simple graph G,

$$t(H, G) \geq t(K_2, G)^{e(H)},$$

where K_2 is the complete graph on two vertices and $e(H)$ is the number of edges in H. Note that $t(K_2, G)$ is simply the edge density of G. The conjecture has been verified in many special cases, such as trees, even cycles, hypercubes and bipartite graphs with one vertex complete to the other part (see Sect. 6.11 for references). The following theorem gives a complete solution to the lower tail problem for bipartite graphs conditional on Sidorenko's conjecture.

Theorem 6.11 *If H is a bipartite graph that satisfies Sidorenko's conjecture, then the lower tail does not have a region of broken symmetry. In particular,* $\phi^-(H, p, t) = I_p(t^{1/e(H)})$ *for all $t \in (0, p^{e(H)})$.*

Proof Suppose that H satisfies Sidorenko's conjecture. Take any $t \in (0, p^{e(H)})$ and let $r = t^{1/e(H)}$. Take any f such that $\tilde{f} \in \widetilde{F}^-(H, p, t)$. By Proposition 3.1 and the Sidorenko property of H,

$$\int_{[0,1]^2} f(x, y)\, dx\, dy = t(K_2, f) \leq t(H, f)^{1/e(H)} \leq r.$$

Since I_p is convex and decreasing in $[0, p]$, the above inequality and Jensen's inequality imply that

$$I_p(f) = \int_{[0,1]^2} I_p(f(x, y))\, dx\, dy \qquad (6.10.1)$$

$$\geq I_p\left(\int_{[0,1]^2} f(x, y)\, dx\, dy\right) \geq I_p(r).$$

Moreover, since I_p is nonlinear everywhere, equality holds if and only if $f = r$ almost everywhere. $\qquad\qquad\square$

6.11   Large Deviations for the Largest Eigenvalue

Let $G_{n,p}$ be a random graph from the Erdős–Rényi $G(n, p)$ model. Let $\lambda_{n,p}$ be the largest eigenvalue of the adjacency matrix of $G_{n,p}$. We have seen in Exercise 6.2 that $\lambda_{n,p}$ equals n times the operator norm of the graphon of $G_{n,p}$. Lemma 6.2 tells us that the operator norm is a nice graph parameter. Therefore the large deviation rate function for the largest eigenvalue and the conditional behavior under large deviations can be obtained by straightforward applications of Theorems 6.1 and 6.2. Explicitly, the result would be that for any $t \in [0, 1]$ (since $[0, 1]$ is the range of the operator norm),

$$\lim_{n\to\infty} \frac{2}{n^2} \mathbb{P}(\lambda_{n,p} \geq tn) = \psi^+(p, t)$$

and

$$\lim_{n\to\infty} \frac{2}{n^2} \mathbb{P}(\lambda_{n,p} \leq tn) = \psi^-(p, t)$$

where

$$\psi^+(p, t) := \inf\{I_p(f) : f \in \mathscr{W}, \|K_f\| \geq t\}$$

and

$$\psi^-(p, t) := \inf\{I_p(f) : f \in \mathscr{W}, \|K_f\| \leq t\}.$$

As for subgraph densities, it is interesting to understand more about the variational problems defining ψ^+ and ψ^-. In particular, one can define the symmetric phase and the region of broken symmetry as before. Note that in this setting, the unique constant optimizer in the symmetric phase is the function that is identically equal to t. If $t \in (p, 1)$ belongs to the symmetric phase, then $\psi^+(p, t) = I_p(t)$, and if $t \in (0, p)$ belongs to the symmetric phase, then $\psi^-(p, t) = I_p(t)$.

Curiously, the phase boundary for the largest eigenvalue turns out to be exactly the same as that of the square root of the homomorphism density of a bipartite graph of degree two (for example, any even cycle). This is the content of the following theorem.

Theorem 6.12 *Take any $p \in (0, 1)$ and $t \in (p, 1)$. If the point $(t^2, I_p(t))$ lies on the convex minorant of the function $J_p(x) := I_p(x^{1/2})$, then t belongs to the symmetric phase of the upper tail in the largest eigenvalue problem. On the other hand, if $(t^2, I_p(t))$ does not lie on the convex minorant of J_p then t belongs to the region of broken symmetry for the upper tail. Lastly, any $t \in (0, p)$ belongs to the symmetric phase of the lower tail.*

Proof Let $\hat{J}_p$ denote the convex minorant of J_p. Take any $t \in (p, 1)$. Suppose that $(t^2, I_p(t))$ lies on the convex minorant of J_p. Take any f such that $\|K_f\| \geq t$. Recall the inequality (3.4.2) from Chap. 3, which says that $\|f\| \geq \|K_f\|$. Therefore, applying the inequality (6.8.1) with $d = 2$ and proceeding as in the proof of Theorem 6.9, we get

$$I_p(f) \geq \hat{J}_p(\|f\|^2) \geq \hat{J}_p(t^2) = J_p(t^2) = I_p(t),$$

with equality if and only if $f = t$ almost everywhere. This proves the first part of the theorem.

Next, suppose that $(t^2, I_p(t))$ does not lie on the convex minorant of J_p. For each $\epsilon > 0$, construct f_ϵ as in the proof of Theorem 6.9, with $d = 2$, and r, r_1 and r_2 replaced by t, t_1 and t_2. Then, as we have seen, $I_p(f_\epsilon) < I_p(t)$ for sufficiently small ϵ. Let a, b, I_0, I_1 and I_2 be as in the proof of Theorem 6.9. Define a function $u : [0, 1] \to \mathbb{R}$ as

$$u(x) := \begin{cases} (1 - a - b)t_1 & \text{if } x \in I_1, \\ (1 - a - b)t_2 & \text{if } x \in I_2, \\ t & \text{if } x \in I_0. \end{cases}$$

If $x \in I_1$, then

$$K_{f_\epsilon} u(x) = a(1 - a - b)t_1 t + b(1 - a - b)t_2 t + (1 - a - b)t_1 t$$
$$> (1 - a - b)t_1 t = tu(x).$$

Similarly, for $x \in I_2$,

$$K_{f_\epsilon} u(x) > (1 - a - b)t_2 t = tu(x).$$

Finally, if $x \in I_0$, then

$$K_{f_\epsilon} u(x) = a(1 - a - b)t_1^2 + b(1 - a - b)t_2^2 + (1 - a - b)t^2$$
$$= (1 - a - b)(t^2 + at_1^2 + bt_2^2).$$

Plugging in the values of a and b the relation $t^2 = st_1^2 + (1 - s)t_2^2$, this gives

$$K_{f_\epsilon} u(x) = t^2 + (t_2^2 - t^2)\epsilon^3 + O(\epsilon^4)$$

as $\epsilon \to 0$, proving that when ϵ is sufficiently small, $K_{f_\epsilon} u(x) > tu(x)$ for all $x \in I_0$. Thus, for ϵ sufficiently small, $K_{f_\epsilon} u(x) > tu(x)$ for almost all $x \in [0, 1]$. Since u and K_{f_ϵ} are nonnegative functions, this implies that $\|K_{f_\epsilon} u\| > t\|u\|$, and therefore $\|K_{f_\epsilon}\| > t$. This proves that t belongs to the region of broken symmetry.

Next, take any $t \in (0, p)$. Take any f such that $\|K_f\| \le t$. Let 1 denote the graphon that is identically equal to 1. Then

$$\int_{[0,1]^2} f(x, y) \, dx \, dy = (1, K_f 1) \le \|1\| \|K_f 1\| \le \|K_f\| \le t.$$

Applying inequality (6.10.1), this gives $I_p(f) \ge I_p(t)$, with equality if and only if $f = t$ almost everywhere, completing the proof of the theorem. □

Bibliographical Notes

The replica symmetric regime for the upper tail of the density of triangles was partially identified in Chatterjee and Dey [1] using techniques based on Stein's method for concentration inequalities. The variational form of the large deviation rate function for triangle density was computed in Chatterjee and Varadhan [2], as an application of the general theory developed in that paper. The double phase transition in the upper tail for triangle density was established in Chatterjee and Varadhan [2], although the identification of the exact phase boundary was left as an open problem. The problem was finally solved by Lubetzky and Zhao [6], in a paper that contains most of the important results presented in this chapter, including Theorem 6.9 (the phase boundary theorem), Theorem 6.10 (double phase transition), Theorem 6.11 (lower tail for Sidorenko graphs) and Theorem 6.12 (large deviations for the largest eigenvalue). The notion of a nice graph parameter was also introduced in Lubetzky and Zhao [6]. A more general investigation of large deviations for random matrices was undertaken in Chatterjee and Varadhan [3].

The Euler–Lagrange equations (Theorem 6.5), the general symmetric phase theorem (Theorem 6.6) and the general symmetry breaking theorem (Theorem 6.7) are new contributions of this monograph.

Sidorenko's conjecture was posed by Erdős and Simonovits in Simonovits [8] and in a slightly more general form by Sidorenko [7]. The conjecture has been proved for trees and even cycles in Sidorenko [7], for hypercubes in Hatami [5] and for bipartite graphs with one vertex complete to the other part in Conlon et al. [4].

One topic that was investigated in Chatterjee and Varadhan [2] but is omitted in this monograph is the behavior of the rate functions when p is sent to zero. The behavior is predictable but the analysis is quite technical, which is the main reason for the omission.

References

1. Chatterjee, S., & Dey, P. S. (2010). Applications of Stein's method for concentration inequalities. *Annals of Probability, 38*, 2443–2485.
2. Chatterjee, S., & Varadhan, S. R. S. (2011). The large deviation principle for the Erdős-Rényi random graph. *European Journal of Combinatorics, 32*(7), 1000–1017.
3. Chatterjee, S., & Varadhan, S. R. S. (2012). Large deviations for random matrices. *Communications on Stochastic Analysis, 6*(1), 1–13.
4. Conlon, D., Fox, J., & Sudakov, B. (2010). An approximate version of Sidorenko's conjecture. *Geometric and Functional Analysis, 20*(6), 1354–1366.
5. Hatami, H. (2010). Graph norms and Sidorenko's conjecture. *Israel Journal of Mathematics, 175*, 125–150.
6. Lubetzky, E. & Zhao, Y. (2015). On replica symmetry of large deviations in random graphs. *Random Structures & Algorithms, 47*(1), 109–146.
7. Sidorenko, A. (1993). A correlation inequality for bipartite graphs. *Graphs and Combinatorics, 9*(2), 201–204.
8. Simonovits, M. (1984). Extremal graph problems, degenerate extremal problems, and supersaturated graphs. In *Progress in graph theory (Waterloo, Ontario, 1982)* (pp. 419–437). Toronto, ON: Academic Press.

Chapter 7
Exponential Random Graph Models

Let $\mathscr{G}_n$ be the space of all simple graphs on n labeled vertices. A variety of probability models on this space can be presented in exponential form

$$p_\beta(G) = \exp\left(\sum_{i=1}^{k} \beta_i T_i(G) - \psi(\beta)\right)$$

where $\beta = (\beta_1, \ldots, \beta_k)$ is a vector of real parameters, $T_1, T_2, \ldots, T_k$ are real-valued functions on $\mathscr{G}_n$, and $\psi(\beta)$ is the normalizing constant. Usually, T_i are taken to be counts of various subgraphs, for example $T_1(G) =$ number of edges in G, $T_2(G) =$ number of triangles in G, etc. These are known as exponential random graph models (ERGM). Our goal in this chapter will be to understand the behavior of random graphs drawn from this class of models, and to calculate the asymptotic values of the normalizing constants. We will continue to use all the notations and terminologies introduced in the preceding chapters.

7.1 Formal Definition Using Graphons

Fix n and let $\mathscr{G}_n$ denote the set of simple graphs on the vertex set $\{1, \ldots, n\}$, as above. For any $G \in \mathscr{G}_n$, recall the definition of the graphon f^G from Chap. 3. Let $\widetilde{f^G}$ be the image of f^G in the quotient space $\widetilde{\mathscr{W}}$. For simplicity, we will write $\widetilde{G}$ instead of $\widetilde{f^G}$.

Let T be a graph parameter, that is, a real-valued continuous function on the space $\widetilde{\mathscr{W}}$. Since $\widetilde{\mathscr{W}}$ is a compact space, T is automatically a bounded function. Define the probability mass function p_n on $\mathscr{G}_n$ induced by T as:

$$p_n(G) := e^{n^2(T(\widetilde{G}) - \psi_n)},$$

© Springer International Publishing AG 2017
S. Chatterjee, *Large Deviations for Random Graphs*, Lecture Notes in Mathematics 2197, DOI 10.1007/978-3-319-65816-2_7

where ψ_n is a constant such that the total mass of p_n is 1. Explicitly,

$$\psi_n = \frac{1}{n^2} \log \sum_{G \in \mathscr{G}_n} e^{n^2 T(\widetilde{G})} \tag{7.1.1}$$

The coefficient n^2 is meant to ensure that ψ_n tends to a non-trivial limit as $n \to \infty$. Note that T does not vary with n.

7.2   Normalizing Constant

Define a function $I : [0, 1] \to \mathbb{R}$ as

$$I(u) := u \log u + (1 - u) \log(1 - u) \tag{7.2.1}$$

and extend I to $\widetilde{\mathscr{W}}$ in the usual manner:

$$I(\widetilde{h}) = \int_{[0,1]^2} I(h(x, y)) \, dx \, dy \tag{7.2.2}$$

where h is a representative element of the equivalence class $\widetilde{h}$. It follows from Proposition 5.1 of Chap. 5 (taking $p = 1/2$) that I is well-defined and lower semi-continuous on $\widetilde{\mathscr{W}}$. The following theorem gives the asymptotic value of the normalizing constant in any exponential random graph model as a variational problem involving the functional I defined above.

Theorem 7.1 *Let T be a graph parameter and ψ_n be the normalizing constant of the exponential random graph model induced by T on the set of simple graphs on n vertices, as defined in Eq. (7.1.1). Let I be the function defined above. Then*

$$\lim_{n \to \infty} \psi_n = \sup_{\widetilde{h} \in \widetilde{\mathscr{W}}} \left(T(\widetilde{h}) - \frac{1}{2} I(\widetilde{h}) \right).$$

Proof For each Borel set $\widetilde{A} \subseteq \widetilde{\mathscr{W}}$ and each n, define

$$\widetilde{A}_n := \{ \widetilde{h} \in \widetilde{A} : \widetilde{h} = \widetilde{G} \text{ for some } G \in \mathscr{G}_n \}.$$

Let $\widetilde{\mathbb{P}}_{n,p}$ be the Erdős–Rényi measure on $\widetilde{\mathscr{W}}$, as defined in Chap. 5. Note that $\widetilde{A}_n$ is a finite set and

$$|\widetilde{A}_n| = 2^{n(n-1)/2} \widetilde{\mathbb{P}}_{n,1/2}(\widetilde{A}_n) = 2^{n(n-1)/2} \widetilde{\mathbb{P}}_{n,1/2}(\widetilde{A}).$$

Thus, if $\widetilde{F}$ is a closed subset of $\widetilde{\mathcal{W}}$, then by Theorem 5.2,

$$\limsup_{n\to\infty} \frac{\log|\widetilde{F}_n|}{n^2} \le \frac{\log 2}{2} - \frac{1}{2}\inf_{\widetilde{h}\in F} I_{1/2}(\widetilde{h})$$

$$= -\frac{1}{2}\inf_{\widetilde{h}\in F} I(\widetilde{h}). \tag{7.2.3}$$

Similarly if $\widetilde{U}$ is an open subset of $\widetilde{\mathcal{W}}$,

$$\liminf_{n\to\infty} \frac{\log|\widetilde{U}_n|}{n^2} \ge -\frac{1}{2}\inf_{\widetilde{h}\in U} I(\widetilde{h}). \tag{7.2.4}$$

Fix $\epsilon > 0$. Since T is a bounded function, there is a finite set R such that the intervals $\{(a, a+\epsilon) : a \in R\}$ cover the range of T. For each $a \in R$, let $\widetilde{F}^a := T^{-1}([a, a+\epsilon])$. By the continuity of T, each $\widetilde{F}^a$ is closed. Now,

$$e^{n^2\psi_n} \le \sum_{a\in R} e^{n^2(a+\epsilon)}|\widetilde{F}^a_n| \le |R| \sup_{a\in R} e^{n^2(a+\epsilon)}|\widetilde{F}^a_n|.$$

By (7.2.3), this shows that

$$\limsup_{n\to\infty} \psi_n \le \sup_{a\in R}\left(a + \epsilon - \frac{1}{2}\inf_{\widetilde{h}\in F^a} I(\widetilde{h})\right).$$

Each $\widetilde{h} \in \widetilde{F}^a$ satisfies $T(\widetilde{h}) \ge a$. Consequently,

$$\sup_{\widetilde{h}\in \widetilde{F}^a}\left(T(\widetilde{h}) - \frac{1}{2}I(\widetilde{h})\right) \ge \sup_{\widetilde{h}\in \widetilde{F}^a}\left(a - \frac{1}{2}I(\widetilde{h})\right) = a - \frac{1}{2}\inf_{\widetilde{h}\in F^a} I(\widetilde{h}).$$

Substituting this in the earlier display gives

$$\limsup_{n\to\infty} \psi_n \le \epsilon + \sup_{a\in R}\sup_{\widetilde{h}\in \widetilde{F}^a}\left(T(\widetilde{h}) - \frac{1}{2}I(\widetilde{h})\right)$$

$$= \epsilon + \sup_{\widetilde{h}\in \widetilde{\mathcal{W}}}\left(T(\widetilde{h}) - \frac{1}{2}I(\widetilde{h})\right). \tag{7.2.5}$$

For each $a \in R$, let $\widetilde{U}^a := T^{-1}((a, a+\epsilon))$. By the continuity of T, $\widetilde{U}^a$ is an open set. Note that

$$e^{n^2\psi_n} \ge \sup_{a\in R} e^{n^2 a}|\widetilde{U}^a_n|.$$

Therefore by (7.2.4), for each $a \in R$

$$\liminf_{n \to \infty} \psi_n \geq a - \frac{1}{2} \inf_{\widetilde{h} \in \widetilde{U}^a} I(\widetilde{h}).$$

Each $\widetilde{h} \in \widetilde{U}^a$ satisfies $T(\widetilde{h}) < a + \epsilon$. Therefore,

$$\sup_{\widetilde{h} \in \widetilde{U}^a} \left(T(\widetilde{h}) - \frac{1}{2} I(\widetilde{h}) \right) \leq \sup_{\widetilde{h} \in \widetilde{U}^a} \left(a + \epsilon - \frac{1}{2} I(\widetilde{h}) \right) = a + \epsilon - \frac{1}{2} \inf_{\widetilde{h} \in \widetilde{U}^a} I(\widetilde{h}).$$

Together with the previous display, this shows that

$$\liminf_{n \to \infty} \psi_n \geq -\epsilon + \sup_{a \in R} \sup_{\widetilde{h} \in \widetilde{U}^a} \left(T(\widetilde{h}) - \frac{1}{2} I(\widetilde{h}) \right)$$

$$= -\epsilon + \sup_{\widetilde{h} \in \mathscr{W}} \left(T(\widetilde{h}) - \frac{1}{2} I(\widetilde{h}) \right). \tag{7.2.6}$$

Since ϵ is arbitrary in (7.2.5) and (7.2.6), this completes the proof.   □

7.3 Asymptotic Structure

Theorem 7.1 gives an asymptotic formula for ψ_n. However, it says nothing about the behavior of a random graph drawn from the exponential random graph model. Some aspects of this behavior can be described as follows. Let $\widetilde{F}^*$ be the subset of $\mathscr{W}$ where $T(\widetilde{h}) - \frac{1}{2} I(\widetilde{h})$ is maximized. By the compactness of $\mathscr{W}$, the continuity of T and the lower semi-continuity of I, $\widetilde{F}^*$ is a non-empty compact set. Let G_n be a random graph on n vertices drawn from the exponential random graph model defined by T. The following theorem shows that for n large, $\widetilde{G}_n$ must lie close to $\widetilde{F}^*$ with high probability. In particular, if $\widetilde{F}^*$ is a singleton set, then the theorem gives a weak law of large numbers for G_n.

Theorem 7.2 *Let $\widetilde{F}^*$ and G_n be defined as in the above paragraph. Let $\mathbb{P}$ denote the probability measure on the underlying probability space on which G_n is defined. Then for any $\eta > 0$ there exist $C, \gamma > 0$ such that for any n,*

$$\mathbb{P}(\delta_\square(\widetilde{G}_n, \widetilde{F}^*) > \eta) \leq C e^{-n^2 \gamma}.$$

Proof Take any $\eta > 0$. Let

$$\widetilde{A} := \{\widetilde{h} : \delta_\square(\widetilde{h}, \widetilde{F}^*) \geq \eta\}.$$

It is easy to see that $\widetilde{A}$ is a closed set. By compactness of $\mathscr{W}$ and $\widetilde{F}^*$, and upper semi-continuity of $T - \frac{1}{2}I$, it follows that

$$2\gamma := \sup_{\widetilde{h} \in \mathscr{W}}\left(T(\widetilde{h}) - \frac{1}{2}I(\widetilde{h})\right) - \sup_{\widetilde{h} \in \widetilde{A}}\left(T(\widetilde{h}) - \frac{1}{2}I(\widetilde{h})\right) > 0.$$

Choose $\epsilon = \gamma$ and define $\widetilde{F}^a$ and R as in the proof of Theorem 7.1. Let $\widetilde{A}^a := \widetilde{A} \cap \widetilde{F}^a$. Then

$$\mathbb{P}(G_n \in \widetilde{A}) \le e^{-n^2\psi_n}\sum_{a \in R}e^{n^2(a+\epsilon)}|\widetilde{A}_n^a| \le e^{-n^2\psi_n}|R|\sup_{a \in R}e^{n^2(a+\epsilon)}|\widetilde{A}_n^a|.$$

While bounding the last term above, it can be assumed without loss of generality that $\widetilde{A}^a$ is non-empty for each $a \in R$, for the other a's can be dropped without upsetting the bound. By (7.2.3) and Theorem 7.1 (noting that $\widetilde{A}^a$ is compact), the above display gives

$$\limsup_{n \to \infty}\frac{\log\mathbb{P}(G_n \in \widetilde{A})}{n^2}$$

$$\le \sup_{a \in R}\left(a + \epsilon - \frac{1}{2}\inf_{\widetilde{h} \in \widetilde{A}^a}I(\widetilde{h})\right) - \sup_{\widetilde{h} \in \mathscr{W}}\left(T(\widetilde{h}) - \frac{1}{2}I(\widetilde{h})\right).$$

Each $\widetilde{h} \in \widetilde{A}^a$ satisfies $T(\widetilde{h}) \ge a$. Consequently,

$$\sup_{\widetilde{h} \in \widetilde{A}^a}\left(T(\widetilde{h}) - \frac{1}{2}I(\widetilde{h})\right) \ge \sup_{\widetilde{h} \in \widetilde{A}^a}\left(a - \frac{1}{2}I(\widetilde{h})\right) = a - \frac{1}{2}\inf_{\widetilde{h} \in \widetilde{A}^a}I(\widetilde{h}).$$

Substituting this in the earlier display gives

$$\limsup_{n \to \infty}\frac{\log\mathbb{P}(G_n \in \widetilde{A})}{n^2} \tag{7.3.1}$$

$$\le \epsilon + \sup_{a \in R}\sup_{\widetilde{h} \in \widetilde{A}^a}\left(T(\widetilde{h}) - \frac{1}{2}I(\widetilde{h})\right) - \sup_{\widetilde{h} \in \mathscr{W}}\left(T(\widetilde{h}) - \frac{1}{2}I(\widetilde{h})\right)$$

$$= \epsilon + \sup_{\widetilde{h} \in \widetilde{A}}\left(T(\widetilde{h}) - \frac{1}{2}I(\widetilde{h})\right) - \sup_{\widetilde{h} \in \mathscr{W}}\left(T(\widetilde{h}) - \frac{1}{2}I(\widetilde{h})\right) = \epsilon - 2\gamma = -\gamma.$$

This completes the proof. $\square$

7.4   An Explicitly Solvable Case

Let $H_1, \ldots, H_k$ be finite simple graphs, where H_1 is the complete graph on two vertices (that is, just a single edge), and each H_i contains at least one edge. Let $\beta_1, \ldots, \beta_k$ be k real numbers. For any $h \in \mathcal{W}$, let

$$T(h) := \sum_{i=1}^{k} \beta_i t(H_i, h) \qquad (7.4.1)$$

where $t(H_i, h)$ is the homomorphism density of H_i in h, defined in Chap. 3. By Proposition 3.2, any such T is a graph parameter. For any finite simple graph G that has at least as many nodes as the largest of the H_i's,

$$T(\widetilde{G}) = \sum_{i=1}^{k} \beta_i t(H_i, G),$$

where $t(H_i, G)$ is the homomorphism density of H_i in G. For example, if $k = 2$, H_2 is a triangle and G has at least three nodes, then

$$T(\widetilde{G}) = 2\beta_1 \frac{\text{number of edges in } G}{n^2} + 6\beta_2 \frac{\text{number of triangles in } G}{n^3}. \qquad (7.4.2)$$

When T is of the form (7.4.1) and $\beta_2, \ldots, \beta_k$ are nonnegative, the following theorem says that the variational problem of Theorem 7.1 can be reduced to a simple maximization problem in one real variable. The theorem moreover says that each solution of the variational problem is a constant function, and there are only a finite number of solutions. By Theorem 7.2, this implies that when $\beta_2, \ldots, \beta_k$ are nonnegative, exponential random graphs from this class of models behave like random graphs drawn from a finite mixture of Erdős–Rényi models.

Theorem 7.3 *Let $H_1, \ldots, H_k$ and T be as above. Suppose that the parameters $\beta_2, \ldots, \beta_k$ are nonnegative. Let ψ_n be the normalizing constant of the exponential random graph model induced by T on the set of simple graphs on n vertices, as defined in Eq. (7.1.1). Then*

$$\lim_{n \to \infty} \psi_n = \sup_{0 \le u \le 1} \left(\sum_{i=1}^{k} \beta_i u^{e(H_i)} - \frac{1}{2} I(u) \right) \qquad (7.4.3)$$

where I is the function defined in (7.2.1) and $e(H_i)$ is the number of edges in H_i. Moreover, there are only a finite number of solutions of the variational problem of Theorem 7.1 for this T, and each solution is a constant function, where the constant solves the scalar maximization problem (7.4.3).

Proof By Theorem 7.1,

$$\lim_{n\to\infty} \psi_n = \sup_{h\in\mathscr{W}}\left(T(h) - \frac{1}{2}I(h)\right). \tag{7.4.4}$$

By Hölder's inequality,

$$t(H_i, h) \le \int_{[0,1]^2} h(x,y)^{e(H_i)}\,dx\,dy.$$

Thus, by the nonnegativity of $\beta_2,\ldots,\beta_k$,

$$T(h) \le \beta_1 t(H_1, h) + \sum_{i=2}^{k} \beta_i \int_{[0,1]^2} h(x,y)^{e(H_i)}\,dx\,dy$$

$$= \int_{[0,1]^2} \sum_{i=1}^{k} \beta_i h(x,y)^{e(H_i)}\,dx\,dy.$$

On the other hand, the inequality in the above display becomes an equality if h is a constant function. Therefore, if u is a point in $[0, 1]$ that maximizes

$$\sum_{i=1}^{k} \beta_i u^{e(H_i)} - \frac{1}{2}I(u),$$

then the constant function $h(x, y) \equiv u$ solves the variational problem (7.4.4). To see that constant functions are the only solutions, assume that there is at least one i such that the graph H_i has at least one vertex with two or more neighbors. The above steps show that if h is a maximizer, then for each i,

$$t(H_i, h) = \int_{[0,1]^2} h(x,y)^{e(H_i)}\,dx\,dy. \tag{7.4.5}$$

In other words, equality holds in Hölder's inequality. Suppose that H_i has vertex set $\{1, 2, \ldots, k\}$ and vertices 2 and 3 are both neighbors of 1 in H_i. Recall that

$$t(H_i, h) = \int_{[0,1]^k} \prod_{\{j,l\}\in E(H_i)} h(x_j, x_l)\,dx_1 \cdots dx_k.$$

In particular, the integrand contains the product $h(x_1, x_2)h(x_1, x_3)$. From this and the criterion for equality in Hölder's inequality, it follows that $h(x_1, x_2)$ is a constant multiple of $h(x_1, x_3)$ for almost every $(x_1, x_2, x_3) \in [0, 1]^3$. Using the symmetry of h one can now easily conclude that h is almost everywhere a constant function.

If the condition does not hold, then each H_i is a union of vertex-disjoint edges. Assume that some H_i has more than one edge. Then again by (7.4.5) it follows that h must be a constant function. Finally, if each H_i is just a single edge, then the maximization problem (7.4.4) can be explicitly solved and the solutions are all constant functions.

Lastly, note that since the second derivative of the map

$$u \mapsto \sum_{i=1}^{k} \beta_i u^{e(H_i)} - \frac{1}{2} I(u)$$

is a rational function of u, Rolle's theorem implies that the set of maximizers is a finite set.   □

7.5   Another Solvable Example

A j-star is an undirected graph with one root vertex and j other vertices connected to the root vertex, with no edges between any of these j vertices. Let H_j be a j-star for $j = 1, \ldots, k$. Let T be the graph parameter defined in (7.4.1) with these $H_1, \ldots, H_k$. It turns out that the exponential random graph model for this T can be explicitly solved for any $\beta_1, \ldots, \beta_k$.

Theorem 7.4 *For the sufficient statistic T defined above, the conclusions of Theorem 7.3 hold for any $\beta_1, \ldots, \beta_k \in \mathbb{R}$.*

Proof Take any $h \in \mathcal{W}$. Note that

$$t(H_j, h) = \int_{[0,1]^j} h(x_1, x_2) h(x_1, x_3) \cdots h(x_1, x_j) \, dx_1 \cdots dx_j$$

$$= \int_0^1 M(x)^j \, dx$$

where

$$M(x) = \int_0^1 h(x, y) \, dy.$$

Since I is a convex function,

$$\int_0^1 I(h(x, y)) \, dy \geq I(M(x)),$$

with equality if and only if $h(x, y)$ is the same for almost all y. Thus, putting

$$P(u) := \sum_{j=1}^{k} \beta_j u^j,$$

we get

$$T(h) - \frac{1}{2}I(h) = \int_0^1 P(M(x)) \, dx - \frac{1}{2}I(h)$$

$$\leq \int_0^1 \left(P(M(x)) - \frac{1}{2}I(M(x)) \right) dx$$

with equality if and only if for almost all x, (a) $h(x, y)$ is constant as a function of y, and (b) $M(x)$ equals a value u^* that maximizes $P(u) - \frac{1}{2}I(u)$. By the symmetry of h, the condition (a) implies that h is constant almost everywhere. The condition (b) gives the set of possible values of this constant. The rest follows as in the proof of Theorem 7.3. □

7.6 Phase Transition in the Edge-Triangle Model

The computations of Sect. 7.4 imply the existence of phase transitions in exponential random graph models. In this section we will demonstrate this through one example. Consider the exponential random graph model corresponding to the graph parameter T defined in (7.4.2). This is sometimes called the 'edge-triangle model'. Let G_n be a random graph on n vertices drawn from this model.

Fix β_1 and β_2 and let

$$\ell(u) := \beta_1 u + \beta_2 u^3 - \frac{1}{2}I(u) \tag{7.6.1}$$

where I is the function defined in (7.2.1). Let U be the set of maximizers of $\ell(u)$ in $[0, 1]$. Theorem 7.3 describes the limiting behavior of G_n in terms of the set U. In particular, if U consists of a single point $u^* = u^*(\beta_1, \beta_2)$, then G_n behaves like the Erdős–Rényi graph $G(n, u^*)$ when n is large.

It is unlikely that $u^*(\beta_1, \beta_2)$ has a closed form expression, other than when $\beta_2 = 0$, in which case

$$u^*(\beta_1, 0) = \frac{e^{2\beta_1}}{1 + e^{2\beta_1}}.$$

We will now present a theorem which shows that for β_1 below a threshold, u^* has a single jump discontinuity in β_2, signifying a first-order phase transition in the parlance of statistical physics.

Theorem 7.5 *Let G_n be a random graph from the edge-triangle model, as defined above. Let*

$$c_1 := \frac{e^{\beta_1}}{1 + e^{\beta_1}}, \quad c_2 := 1 + \frac{1}{2\beta_1}.$$

Suppose that $\beta_1 < 0$ and $|\beta_1|$ is so large that $c_1 < c_2$. Let $e(G_n)$ be the number of edges in G_n and let $f(G_n) := e(G_n)/\binom{n}{2}$ be the edge density. Then there exists $q = q(\beta_1) \in [0, \infty)$ such that if $-\infty < \beta_2 < q$, then

$$\lim_{n \to \infty} \mathbb{P}(f(G_n) > c_1) = 0,$$

and if $\beta_2 > q$, then

$$\lim_{n \to \infty} \mathbb{P}(f(G_n) < c_2) = 0.$$

Proof Fix $\beta_1 < 0$ such that $c_1 < c_2$. As a preliminary step, let us prove that for any $\beta_2 > 0$,

$$\lim_{n \to \infty} \mathbb{P}(f(G_n) \in (c_1, c_2)) = 0. \tag{7.6.2}$$

Fix $\beta_2 > 0$. Let u be any maximizer of the function ℓ defined in (7.6.1). Then by Theorem 7.3, it suffices to prove that either $u < e^{\beta_1}/(1 + e^{\beta_1})$ or $u > 1 + 1/2\beta_1$. This is proved as follows. Define a function $g : [0, 1] \to \mathbb{R}$ as

$$g(v) := \ell(v^{1/3}).$$

Then ℓ is maximized at u if and only if g is maximized at u^3. Since ℓ is a bounded continuous function and $\ell'(0) = \infty$ and $\ell'(1) = -\infty$, ℓ cannot be maximized at 0 or 1. Therefore the same is true for g. Let v be a point in $(0, 1)$ at which g is maximized. Then $g''(v) \le 0$. A simple computation shows that

$$g''(v) = \frac{1}{9v^{5/3}}\left(-2\beta_1 + \log \frac{v^{1/3}}{1 - v^{1/3}} - \frac{1}{2(1 - v^{1/3})}\right).$$

Thus, $g''(v) \le 0$ only if

$$\log \frac{v^{1/3}}{1 - v^{1/3}} \le \beta_1 \quad \text{or} \quad -\frac{1}{2(1 - v^{1/3})} \le \beta_1.$$

This shows that a maximizer u of ℓ must satisfy $u \leq c_1$ or $u \geq c_2$. Now, if $u = c_1$, then $u < c_2$, and therefore the above computations show that $g''(v) > 0$, where $v = u^3$. Similarly, if $u = c_2$ then $u > c_1$ and again $g''(v) > 0$. Thus, we have proved that $u < c_1$ or $u > c_2$. By Theorem 7.2, this completes the proof of (7.6.2) when $\beta_2 > 0$.

Now notice that as $\beta_2 \to \infty$, $\sup_{u \leq a} \ell(u) \sim \beta_2 a^3$ for any fixed $a \leq 1$. This shows that as $\beta_2 \to \infty$, any maximizer of ℓ must eventually be larger than $1 + 1/2\beta_1$. Therefore, for sufficiently large β_2,

$$\lim_{n \to \infty} \mathbb{P}(f(G_n) < c_2) = 0. \tag{7.6.3}$$

Next consider the case $\beta_2 \leq 0$. Let $\widetilde{F}^*$ be the set of maximizers of $T(\widetilde{h}) - \frac{1}{2}I(\widetilde{h})$. Take any $\widetilde{h} \in \widetilde{F}^*$ and let h be a representative element of $\widetilde{h}$. Let $p = e^{2\beta_1}/(1 + e^{2\beta_1})$. An easy verification shows that

$$T(h) - \frac{1}{2}I(h) = \beta_2 t(H_2, h) - \frac{1}{2}I_p(h) - \frac{1}{2}\log(1 - p),$$

where $I_p(h)$ is defined in Eq. (5.2.1) of Chap. 5. Define a new function

$$h_1(x, y) := \min\{h(x, y), p\}.$$

Since the function I_p defined in (5.2.1) is minimized at p, it follows that for all $x, y \in [0, 1]$, $I_p(h_1(x, y)) \leq I_p(h(x, y))$. Consequently, $I_p(h_1) \leq I_p(h)$. Again, since $\beta_2 \leq 0$ and $h_1 \leq h$ everywhere, $\beta_2 t(H_2, h_1) \geq \beta_2 t(H_2, h)$. Combining these observations, we see that $T(h_1) - \frac{1}{2}I(h_1) \geq T(h) - \frac{1}{2}I(h)$. Since h maximizes $T - \frac{1}{2}I$ it follows that equality must hold at every step in the above deductions, from which it is easy to conclude that $h = h_1$ a.e. In other words, $h(x, y) \leq p$ a.e. This is true for every $\widetilde{h} \in \widetilde{F}^*$. Since $p < c_1$, the above deduction coupled with Theorem 7.2 proves that when $\beta_2 \leq 0$,

$$\lim_{n \to \infty} \mathbb{P}(f(G_n) > c_1) = 0. \tag{7.6.4}$$

Recalling that β_1 is fixed, define

$$a_n(\beta_2) := \mathbb{P}(f(G_n) > c_1), \quad b_n(\beta_2) := \mathbb{P}(f(G_n) < c_2).$$

Let A_n and B_n denote the events in brackets in the above display. A simple computation shows that

$$a_n'(\beta_2) = \frac{6}{n}\text{Cov}(1_{A_n}, \Delta(G_n)) \quad \text{and} \quad b_n'(\beta_2) = \frac{6}{n}\text{Cov}(1_{B_n}, \Delta(G_n)),$$

where $\Delta(G_n)$ is the number of triangles in G_n. It is easy to verify that the edge-triangle model with $\beta_2 \geq 0$ satisfies the FKG lattice criterion (2.6.1) stated in Chap. 2. Moreover, 1_{A_n} and Δ are increasing functions of the edge variables, and 1_{B_n} is a decreasing function. Therefore the above identities and the FKG inequality (Theorem 2.2 of Chap. 2) show that on the nonnegative axis, a_n is a non-decreasing function and b_n is a non-increasing function.

Let $q_1 := \sup\{x \in \mathbb{R} : \lim_{n \to \infty} a_n(x) = 0\}$. By Eq. (7.6.3), $q_1 < \infty$ and by Eq. (7.6.4) $q_1 \geq 0$. Similarly, if $q_2 := \inf\{x \in \mathbb{R} : \lim_{n \to \infty} b_n(x) = 0\}$, then $0 \leq q_2 < \infty$. Also, clearly, $q_1 \leq q_2$ since $a_n + b_n \geq 1$ everywhere. We claim that $q_1 = q_2$. This would complete the proof by the monotonicity of a_n and b_n.

To prove that $q_1 = q_2$, suppose not. Then $q_1 < q_2$. Then for any $\beta_2 \in (q_1, q_2)$, $\limsup a_n(\beta_2) > 0$ and $\limsup b_n(\beta_2) > 0$. Now,

$$0 \leq a_n(\beta_2) + b_n(\beta_2) - 1 = \mathbb{P}(f(G_n) \in (c_1, c_2)).$$

Therefore by (7.6.2),

$$\lim_{n \to \infty} (a_n(\beta_2) + b_n(\beta_2) - 1) = 0.$$

Thus, for any $\beta_2 \in (q_1, q_2)$, $\limsup(1 - b_n(\beta_2)) > 0$. By Theorem 7.3, this implies that the function ℓ has a maximum in $[c_2, 1]$. Similarly, for any $\beta_2 \in (q_1, q_2)$, $\limsup(1 - a_n(\beta_2)) > 0$ and therefore the function ℓ has a maximum in $[0, c_1]$. Now fix $q_1 < \beta_2 < \tilde{\beta}_2 < q_2$, and let ℓ and $\tilde{\ell}$ denote the two ℓ-functions corresponding to β_2 and $\tilde{\beta}_2$ respectively. That is,

$$\ell(u) = \beta_1 u + \beta_2 u^3 - \frac{1}{2}I(u), \quad \tilde{\ell}(u) = \beta_1 u + \tilde{\beta}_2 u^3 - \frac{1}{2}I(u).$$

By the above argument, ℓ attains its maximum at some point $u_1 \in [0, c_1]$ and at some point $u_2 \in [c_2, 1]$. (There may be other maxima, but that is irrelevant for us.) Note that

$$\max_{u \leq c_1} \tilde{\ell}(u) = \max_{u \leq c_1}(\ell(u) + (\tilde{\beta}_2 - \beta_2)u^3) \leq \ell(u_1) + (\tilde{\beta}_2 - \beta_2)c_1^3.$$

On the other hand

$$\max_{u \geq c_2} \tilde{\ell}(u) \geq \tilde{\ell}(u_2) = \ell(u_2) + (\tilde{\beta}_2 - \beta_2)u_2^3 \geq \ell(u_2) + (\tilde{\beta}_2 - \beta_2)c_2^3.$$

Since $\ell(u_1) = \ell(u_2)$, $\tilde{\beta}_2 > \beta_2$ and $c_2 > c_1$, this shows that

$$\max_{u \leq c_1} \tilde{\ell}(u) < \max_{u \geq c_2} \tilde{\ell}(u),$$

contradicting our previous deduction that $\tilde{\ell}$ has maxima in both $[0, c_1]$ and $[c_2, 1]$. This proves that $q_1 = q_2$. $\square$

7.7   Euler–Lagrange Equations

In this section we will derive the Euler–Lagrange equation for the solution of the variational problem of Theorem 7.1. The analysis is very similar to the one carried out in Sect. 6.5 of Chap. 6. Let the operator Δ_H be defined as in Eq. (6.5.1) of that section.

Theorem 7.6 *Let T be as in Eq. (7.4.1). If $h \in \mathcal{W}$ maximizes $T(h) - \frac{1}{2}I(h)$, then for almost all $(x, y) \in [0, 1]^2$,*

$$h(x, y) = \frac{e^{2\sum_{i=1}^{k} \beta_i \Delta_{H_i} h(x,y)}}{1 + e^{2\sum_{i=1}^{k} \beta_i \Delta_{H_i} h(x,y)}}.$$

Moreover, h is bounded away from 0 and 1.

Proof Let g be a symmetric bounded measurable function from $[0, 1]$ into $\mathbb{R}$. For each $u \in \mathbb{R}$, let

$$h_u(x, y) := h(x, y) + ug(x, y).$$

Then h_u is a symmetric bounded measurable function from $[0, 1]$ into $\mathbb{R}$. First suppose that h is bounded away from 0 to 1. Then $h_u \in \mathcal{W}$ for every u sufficiently small in magnitude. Since h maximizes $T(h) - \frac{1}{2}I(h)$ among all elements of $\mathcal{W}$, therefore under the above assumption, for all u sufficiently close to zero,

$$T(h_u) - \frac{1}{2}I(h_u) \le T(h) - \frac{1}{2}I(h).$$

In particular,

$$\frac{d}{du}\left(T(h_u) - \frac{1}{2}I(h_u)\right)\bigg|_{u=0} = 0. \tag{7.7.1}$$

It is easy to check that $T(h_u) - \frac{1}{2}I(h_u)$ is differentiable in u for any h and g. In particular, the derivative is given by

$$\frac{d}{du}\left(T(h_u) - \frac{1}{2}I(h_u)\right) = \sum_{i=1}^{k} \beta_i \frac{d}{du} t(H_i, h_u) - \frac{1}{2}\frac{d}{du}I(h_u).$$

Now,

$$\frac{1}{2}\frac{d}{du}I(h_u) = \frac{1}{2}\int_{[0,1]^2} \frac{d}{du}I(h(x, y) + ug(x, y))\, dy\, dx$$

$$= \frac{1}{2}\int_{[0,1]^2} g(x, y) \log \frac{h_u(x, y)}{1 - h_u(x, y)}\, dy\, dx.$$

Consequently,

$$\frac{1}{2}\frac{d}{du}I(h_u)\Big|_{u=0} = \frac{1}{2}\int_{[0,1]^2} g(x,y) \log \frac{h(x,y)}{1-h(x,y)} \, dy \, dx.$$

Next, note that

$$\frac{d}{du}t(H_i, h_u)$$

$$= \int_{[0,1]^{V(H)}} \sum_{(r,s)\in E(H_i)} g(x_r, x_s) \prod_{\substack{\{r',s'\}\in E(H_i) \\ \{r',s'\}\neq\{r,s\}}} h_u(x_{r'}, x_{s'}) \prod_{v\in V(H_i)} dx_v$$

$$= \int_{[0,1]^2} g(x,y)\Delta_{H_i}h_u(x,y) \, dy \, dx.$$

Combining the above computations and (7.7.1), we see that for any symmetric bounded measurable $g : [0,1] \to \mathbb{R}$,

$$\iint g(x,y)\left(\sum_{i=1}^{k} \beta_i\Delta_{H_i}h(x,y) - \frac{1}{2}\log\frac{h(x,y)}{1-h(x,y)}\right) dy \, dx = 0.$$

Taking $g(x,y)$ equal to the function within the brackets (which is bounded since h is assumed to be bounded away from 0 and 1), the conclusion of the theorem follows.

Now recall that the theorem was proved under the assumption that h is bounded away from 0 and 1. We claim that this is true for any h that maximizes $T(h) - \frac{1}{2}I(h)$. To prove this claim, take any such h. Fix $p \in (0, 1)$. For each $u \in [0, 1]$, let

$$h_{p,u}(x,y) := h(x,y) + u(p - h(x,y))_+.$$

In other words, $h_{p,u}$ is simply h_u with $g = (p - h)_+$. Then $h_{p,u}$ is a symmetric bounded measurable function from $[0,1]^2$ into $[0,1]$. Note that

$$\frac{d}{du}h_{p,u}(x,y) = (p - h(x,y))_+.$$

Using this, an easy computation as above shows that

$$\frac{d}{du}\left(T(h_{p,u}) - \frac{1}{2}I(h_{p,u})\right)\Big|_{u=0}$$

$$= \int_{[0,1]^2} \left(\sum_{i=1}^{k} \beta_i\Delta_{H_i}h(x,y) - \frac{1}{2}\log\frac{h(x,y)}{1-h(x,y)}\right)(p - h(x,y))_+ \, dy \, dx$$

$$\geq \int_{[0,1]^2} \left(-C - \frac{1}{2}\log\frac{h(x,y)}{1-h(x,y)}\right)(p - h(x,y))_+ \, dy \, dx$$

where C is a positive constant depending only on $\beta_1, \ldots, \beta_k$ and $H_1, \ldots, H_k$ (and not on p or h). When $h(x, y) = 0$, the integrand is interpreted as ∞, and when $h(x, y) = 1$, the integrand is interpreted as 0.

Now, if p is so small that

$$-C - \frac{1}{2} \log \frac{p}{1-p} > 0,$$

then the previous display proves that the derivative of $T(h_{p,u}) - \frac{1}{2}I(h_{p,u})$ with respect to u is strictly positive at $u = 0$ if $h < p$ on a set of positive Lebesgue measure. Hence h cannot be a maximizer of $T - \frac{1}{2}I$ unless $h \geq p$ almost everywhere. This proves that any maximizer of $T - \frac{1}{2}I$ must be bounded away from zero. A similar argument with $g = -(h - p)_+$ shows that it must be bounded away from 1 and hence completes the proof of the theorem.   □

7.8 The Symmetric Phase

This section is the analog of Sect. 6.6 of Chap. 6 for exponential random graph models. The following theorem shows that for the statistic T defined in Eq. (7.4.1), there is a unique maximizer of $T(h) - \frac{1}{2}I(h)$ if $|\beta_2|, \ldots, |\beta_k|$ are small enough. The proof involves a simple application of Euler–Lagrange equation presented in Theorem 7.6. Recall that the same statement was proved in Theorem 7.3 under the hypothesis that $\beta_2, \ldots, \beta_k$ are nonnegative.

Theorem 7.7 *Let T be as in Eq. (7.4.1). Suppose $\beta_1, \ldots, \beta_k$ are such that*

$$\sum_{i=2}^{k} |\beta_i| e(H_i)(e(H_i) - 1) < 2$$

where $e(H_i)$ is the number of edges in H_i. Then the conclusions of Theorem 7.3 hold.

Proof It suffices to prove that the maximizer of $T - \frac{1}{2}I$ is unique. This is because if h is a maximizer, then so is $h_\sigma(x, y) := h(\sigma x, \sigma y)$ for any measure preserving bijection $\sigma : [0, 1] \to [0, 1]$. The only functions that are invariant under such transforms are functions that are constant almost everywhere.

Let Δ_H be the operator defined in Sect. 6.5 of Chap. 6. Let $\| \cdot \|_\infty$ denote the L^∞ norm on $\mathcal{W}$. Let h and g be two maximizers of $T - \frac{1}{2}I$. For any finite simple graph H, a simple computation shows that

$$\| \Delta_H h - \Delta_H g \|_\infty \leq \sum_{(r,s) \in E(H)} \| \Delta_{H,r,s} h - \Delta_{H,r,s} g \|_\infty$$

$$\leq e(H)(e(H) - 1) \| h - g \|_\infty.$$

Using the above inequality, Theorem 7.6 and the inequality

$$\left| \frac{e^x}{1+e^x} - \frac{e^y}{1+e^y} \right| \leq \frac{|x-y|}{4},$$

it follows that for almost all x, y,

$$|h(x,y) - g(x,y)| = \left| \frac{e^{2\sum_{i=1}^{k} \beta_i \Delta_{H_i} h(x,y)}}{1 + e^{2\sum_{i=1}^{k} \beta_i \Delta_{H_i} h(x,y)}} - \frac{e^{2\sum_{i=1}^{k} \beta_i \Delta_{H_i} g(x,y)}}{1 + e^{2\sum_{i=1}^{k} \beta_i \Delta_{H_i} g(x,y)}} \right|$$

$$\leq \frac{1}{2} \sum_{i=1}^{k} |\beta_i| \|\Delta_{H_i} h - \Delta_{H_i} g\|_\infty$$

$$\leq \frac{1}{2} \|h - g\|_\infty \sum_{i=1}^{k} |\beta_i| e(H_i)(e(H_i) - 1).$$

If the coefficient of $\|h - g\|_\infty$ in the last expression is strictly less than 1, it follows that h must be equal to g almost everywhere. $\qquad\square$

7.9 Symmetry Breaking

Let T be as in (7.4.1). Theorems 7.3 and 7.7 give sufficient conditions for $T - \frac{1}{2}I$ to be maximized by a constant function. Is it possible that $T - \frac{1}{2}I$ has a non-constant maximizer in certain situations? In analogy with Sect. 6.7 of Chap. 6, we call this 'symmetry breaking'. The following theorem shows that symmetry breaking is indeed possible, by considering the specific case of the edge-triangle model. As discussed in Sect. 6.7, symmetry breaking implies that a typical random graph from the model does not 'look like' an Erdős–Rényi graph.

Theorem 7.8 *Let T be as in Eq. (7.4.2). Then for any given value of β_1, there is a positive constant $C(\beta_1)$ sufficiently large so that whenever $\beta_2 < -C(\beta_1)$, $T - \frac{1}{2}I$ is not maximized at any constant function.*

Proof Fix β_1. Let $p = e^{2\beta_1}/(1 + e^{2\beta_1})$ and $\gamma = -\beta_2$, so that for any $h \in \mathcal{W}$,

$$T(h) - \frac{1}{2}I(h) = -\gamma t(H_2, h) - \frac{1}{2}I_p(h) - \frac{1}{2}\log(1 - p).$$

Assume without loss of generality that $\beta_2 < 0$. Suppose that u is a constant such that the constant function $h(x, y) \equiv u$ maximizes $T(h) - \frac{1}{2}I(h)$, that is, minimizes $\gamma t(H_2, h) + \frac{1}{2}I_p(h)$. Note that

$$\gamma t(H_2, h) + \frac{1}{2}I_p(h) = \gamma u^3 + \frac{1}{2}I_p(u).$$

Clearly, the definition of u implies that $\gamma u^3 + \frac{1}{2}I_p(u) \leq \gamma x^3 + \frac{1}{2}I_p(x)$ for all $x \in [0, 1]$. This implies that u must be in $(0, 1)$, because the derivative of $x \mapsto \gamma x^3 + \frac{1}{2}I_p(x)$ is $-\infty$ at 0 and ∞ at 1. Thus,

$$0 = \frac{d}{dx}\left(\gamma x^3 + \frac{1}{2}I_p(x)\right)\bigg|_{x=u} = 3\gamma u^2 + \frac{1}{2}\log\frac{u}{1-u} - \frac{1}{2}\log\frac{p}{1-p}$$

which shows that $u \leq c(\gamma)$, where $c(\gamma)$ is a function of γ such that

$$\lim_{\gamma \to \infty} c(\gamma) = 0.$$

This shows that

$$\lim_{\gamma \to \infty} \min_{0 \leq x \leq 1}\left(\gamma x^3 + \frac{1}{2}I_p(x)\right) = \frac{1}{2}I_p(0) = \frac{1}{2}\log\frac{1}{1-p}. \tag{7.9.1}$$

Next let g be the function

$$g(x, y) := \begin{cases} 0 & \text{if } x, y \text{ on same side of } 1/2 \\ p & \text{if not.} \end{cases}$$

Clearly, for almost all $(x, y, z) \in [0, 1]^3$, $g(x, y)g(y, z)g(z, x) = 0$. Thus, $t(H_2, g) = 0$. A simple computation shows that

$$\frac{1}{2}I_p(g) = \frac{1}{4}\log\frac{1}{1-p}.$$

Therefore $\gamma t(H_2, g) + \frac{1}{2}I_p(g) = \frac{1}{4}\log\frac{1}{1-p}$. This shows that if γ is large enough (depending on p and hence β_1), then $T - \frac{1}{2}I$ cannot be maximized at a constant function.   □

Bibliographical Notes

Exponential random graph models have a long history in the statistics literature. They were used by Holland and Leinhardt [10] in the directed case. Frank and Strauss [7] developed them, showing that if T_i are chosen as edges, triangles and stars of various sizes, the resulting random graph edges form a Markov random field. A general development is in Wasserman and Faust [20]. Newer developments, consisting mainly of new sufficient statistics and new ranges for parameters that give interesting and practically relevant structures, are summarized in Snijders et al. [19]. Rinaldo et al. [18] developed the geometric theory for this class of models with extensive further references.

A major problem in this field is the evaluation of the normalizing constant, which is crucial for carrying out maximum likelihood and Bayesian inference. For a long time, there used to exist no feasible analytic method for approximating the normalizing constant when n is large. Physicists had tried the unrigorous technique of mean field approximations; see Park and Newman [14, 15] for the case where T_1 is the number of edges and T_2 is the number of two-stars or the number of triangles. For exponential graph models, Chatterjee and Dey [4] proved that these approximations work in certain subregions of the symmetric phase. This was further developed in Bhamidi et al. [2]. The problem of analytically computing the normalizing constant in exponential random graph models was finally solved in Chatterjee and Diaconis [5] using the large deviation results of Chatterjee and Varadhan [6]. The theorems presented in this chapter are all reproduced from Chatterjee and Diaconis [5].

Recently, the mean field approach for analyzing exponential random graph models was made fully rigorous in Chatterjee and Dembo [3]. The results of this chapter could also have been derived using the methods of Chatterjee and Dembo [3] instead of graph limit theory. This is further discussed in Chap. 8.

The phase transition discussed in Sect. 7.6 is a mathematically precise version of the phenomenon of 'near-degeneracy' that was observed and studied in Handcock [9] (see also Park and Newman [14] and Häggström and Jonasson [8]). Near-degeneracy, in this context, means that when β_1 is a large negative number, then as β_2 varies, the model transitions from being a very sparse graph for low values of β_2, to a very dense graph for large values of β_2, completely skipping all intermediate structures. Significant extensions of Theorem 7.5 have been made in Aristoff and Radin [1], Radin and Sadun [16], Radin and Yin [17] and Yin [21].

The symmetry breaking phenomenon discussed in Sect. 7.9 has been further investigated by Lubetzky and Zhao [13]. They showed that if in the edge-triangle model, the homomorphism density of triangles is replaced by the homomorphism density raised to the power α for some $\alpha \in (0, 2/3)$, then the model can exhibit symmetry breaking even if β_2 is nonnegative.

As far as I know, there is no T for which a non-constant optimizer of $T - \frac{1}{2}I$ has been explicitly computed, either analytically or using a computer. There is however, one related class of models—called constrained graph models—where a similar theory has been developed and non-constant optimizers have been explicitly computed. See Kenyon et al. [12] and Kenyon and Yin [11] for details.

References

1. Aristoff, D., & Radin, C. (2013). Emergent structures in large networks. *Journal of Applied Probability, 50*(3), 883–888.
2. Bhamidi, S., Bresler, G., & Sly, A. (2008). Mixing time of exponential random graphs. In *2008 IEEE 49th Annual IEEE Symposium on Foundations of Computer Science (FOCS)* (pp. 803–12).

3. Chatterjee, S., & Dembo, A. (2016). Nonlinear large deviations. *Advances in Mathematics, 299*, 396–450.
4. Chatterjee, S., & Dey, P. S. (2010). Applications of Stein's method for concentration inequalities. *Annals of Probability, 38*, 2443–2485.
5. Chatterjee, S., & Diaconis, P. (2013). Estimating and understanding exponential random graph models. *Annals of Statistics, 41*(5), 2428–2461.
6. Chatterjee, S., & Varadhan, S. R. S. (2011). The large deviation principle for the Erdős-Rényi random graph. *European Journal of Combinatorics, 32*(7), 1000–1017.
7. Frank, O., & Strauss, D. (1986). Markov graphs. *Journal of the American Statistical Association, 81*, 832–842
8. Häggström, O., & Jonasson, J. (1999). Phase transition in the random triangle model. *Journal of Applied Probability, 36*(4), 1101–1115.
9. Handcock, M. S. (2003). *Assessing degeneracy in statistical models of social networks, Working Paper 39*. Technical report. Center for Statistics and the Social Sciences, University of Washington.
10. Holland, P. W., & Leinhardt, S. (1981). An exponential family of probability distributions for directed graphs. *Journal of the American Statistical Association, 76*(373), 33–65.
11. Kenyon, R., & Yin, M. (2014). On the asymptotics of constrained exponential random graphs. *arXiv preprint arXiv:1406.3662*.
12. Kenyon, R., Radin, C., Ren, K., & Sadun, L. (2014). Multipodal structure and phase transitions in large constrained graphs. *arXiv preprint arXiv:1405.0599*.
13. Lubetzky, E. & Zhao, Y. (2015). On replica symmetry of large deviations in random graphs. *Random Structures & Algorithms, 47*(1), 109–146.
14. Park, J., & Newman, M. E. J. (2004). Solution of the two-star model of a network. *Physical Review E (3), 70*, 066146, 5.
15. Park, J., & Newman, M. E. J. (2005). Solution for the properties of a clustered network. *Physical Review E (3), 72*, 026136, 5.
16. Radin, C., & Sadun, L. (2013). Phase transitions in a complex network. *Journal of Physics A, 46*, 305002.
17. Radin, C., & Yin, M. (2011). Phase transitions in exponential random graphs. *arXiv preprint arXiv:1108.0649*.
18. Rinaldo, A., Fienberg, S. E., & Zhou, Y. (2009). On the geometry of discrete exponential families with application to exponential random graph models. *Electronic Journal of Statistics, 3*, 446–484.
19. Snijders, T. A. B., Pattison, P. E., Robins, G. L., & Handcock, M. S. (2006). New specifications for exponential random graph models. *Sociological Methodology, 36*, 99–153.
20. Wasserman, S., & Faust, K. (2010). *Social network analysis: Methods and applications* (2nd ed.). In, Structural Analysis in the Social Sciences. Cambridge: Cambridge University Press.
21. Yin, M. (2013). Critical phenomena in exponential random graphs. *Journal of Statistical Physics, 153*(6), 1008–1021.

Chapter 8
Large Deviations for Sparse Graphs

The development in this monograph till now has been based on results from graph limit theory. This theory, however, is inadequate for understanding the behavior of sparse graphs. The goal of this chapter is to describe an alternative approach, called nonlinear large deviations, that allows us to prove similar results for sparse graphs. Nonlinear large deviation theory gives a way of getting quantitative error bounds in some of the large deviation theorems proved in earlier chapters. The quantitative error bounds make it possible to extend the results to the sparse regime. At the time of writing this monograph, this theory is not as well-developed as the theory for dense graphs, but developed enough to make some progress about sparse graphs.

8.1 An Approximation for Partition Functions

Take any $N \geq 1$. Let $\|f\|$ denote the supremum norm of any function $f : [0, 1]^N \to \mathbb{R}$. Suppose that $f : [0, 1]^N \to \mathbb{R}$ is twice continuously differentiable in $(0, 1)^N$, such that f and all its first and second order derivatives extend continuously to the boundary. For each i and j, let

$$f_i := \frac{\partial f}{\partial x_i} \quad \text{and} \quad f_{ij} := \frac{\partial^2 f}{\partial x_i \partial x_j}.$$

Define

$$a := \|f\|, \quad b_i := \|f_i\| \quad \text{and} \quad c_{ij} := \|f_{ij}\|.$$

Given $\epsilon > 0$, let $\mathscr{D}(\epsilon)$ be a finite subset of $\mathbb{R}^N$ such that for all $x \in \{0, 1\}^N$, there exists $d = (d_1, \ldots, d_N) \in \mathscr{D}(\epsilon)$ such that

$$\sum_{i=1}^{N} (f_i(x) - d_i)^2 \leq N\epsilon^2. \tag{8.1.1}$$

© Springer International Publishing AG 2017
S. Chatterjee, *Large Deviations for Random Graphs*, Lecture Notes in Mathematics 2197, DOI 10.1007/978-3-319-65816-2_8

Define

$$F := \log \sum_{x \in \{0,1\}^N} e^{f(x)}.$$

In the terminology of statistical mechanics, F is the logarithm of the partition function of the probability measure on $\{0, 1\}^N$ with Hamiltonian f. For $x \in [0, 1]$, let

$$I(x) := x \log x + (1-x) \log(1-x), \tag{8.1.2}$$

and for $x = (x_1, \ldots, x_N) \in [0, 1]^N$, let

$$I(x) := \sum_{i=1}^{N} I(x_i).$$

The following theorem gives a sufficient condition on f under which the approximation

$$F = \sup_{x \in [0,1]^N} (f(x) - I(x)) + \text{lower order terms} \tag{8.1.3}$$

is valid. This is sometimes called the 'naive mean field approximation' in the statistical physics literature. Roughly speaking, the condition is that in addition to some smoothness assumptions, the gradient vector $\nabla f(x) = (\partial f / \partial x_1, \ldots, \partial f / \partial x_N)$ may be approximately encoded by $o(N)$ bits of information. A bit more precisely, we need $|\mathscr{D}(\epsilon)| = e^{o(N)}$ for some $\epsilon = o(1)$, where the implicit assumption is that $N \to \infty$ and f varies with N. We will refer to this as the 'low complexity gradient' condition.

Theorem 8.1 *Let f, a, b_i, c_{ij}, $\mathscr{D}(\epsilon)$, F and I be as above. Then for any $\epsilon > 0$,*

$$F \leq \sup_{x \in [0,1]^N} (f(x) - I(x)) + \text{complexity} + \text{smoothness},$$

where

$$\text{complexity} = \frac{1}{4} \left(N \sum_{i=1}^{N} b_i^2 \right)^{1/2} \epsilon + 3N\epsilon + \log |\mathscr{D}(\epsilon)|, \quad \text{and}$$

$$\text{smoothness} = 4 \left(\sum_{i=1}^{N} (ac_{ii} + b_i^2) + \frac{1}{4} \sum_{i,j=1}^{N} (ac_{ij}^2 + b_i b_j c_{ij} + 4 b_i c_{ij}) \right)^{1/2}$$

$$+ \frac{1}{4} \left(\sum_{i=1}^{N} b_i^2 \right)^{1/2} \left(\sum_{i=1}^{N} c_{ii}^2 \right)^{1/2} + 3 \sum_{i=1}^{N} c_{ii} + \log 2.$$

Moreover, F satisfies the lower bound

$$F \geq \sup_{x \in [0,1]^N} (f(x) - I(x)) - \frac{1}{2} \sum_{i=1}^{N} c_{ii}.$$

The rest of this section is devoted to the proof of Theorem 8.1. Examples are worked in subsequent sections.

We will generally denote the ith coordinate of a vector $x \in \mathbb{R}^N$ by x_i. Similarly, the ith coordinate of a random vector X will be denoted by X_i. Given $x \in [0,1]^N$, define $x^{(i)}$ to be the vector $(x_1, \ldots, x_{i-1}, 0, x_{i+1}, \ldots, x_N)$. For a random vector X define $X^{(i)}$ similarly. Given a function $g : [0,1]^N \to \mathbb{R}$, define the discrete derivative $\Delta_i g$ as

$$\Delta_i g(x) := g(x_1, \ldots, x_{i-1}, 1, x_{i+1}, \ldots, x_N)$$
$$- g(x_1, \ldots, x_{i-1}, 0, x_{i+1}, \ldots, x_N). \tag{8.1.4}$$

For each i, define a function $\hat{x}_i : [0,1]^N \to [0,1]$ as

$$\hat{x}_i(x) = \frac{1}{1 + e^{-\Delta_i f(x)}}.$$

Let $\hat{x} : [0,1]^N \to [0,1]^N$ be the vector-valued function whose ith coordinate function is $\hat{x}_i$. When the vector x is understood from the context, we will simply write $\hat{x}$ and $\hat{x}_i$ instead of $\hat{x}(x)$ and $\hat{x}_i(x)$. The proof of Theorem 8.1 requires two key lemmas.

Lemma 8.1 *Let $X = (X_1, \ldots, X_N)$ be a random vector that has probability density proportional to $e^{f(x)}$ on $\{0,1\}^N$ with respect to the counting measure. Let $\hat{X} = \hat{x}(X)$. Then*

$$\mathbb{E}\big[(f(X) - f(\hat{X}))^2\big] \leq \sum_{i=1}^{N}(ac_{ii} + b_i^2) + \frac{1}{4} \sum_{i,j=1}^{N}\big(ac_{ij}^2 + b_i b_j c_{ij}\big).$$

Proof It is easy to see that

$$\hat{x}_i(x) = \mathbb{E}(X_i \mid X_j = x_j, \ 1 \leq j \leq N, \ j \neq i).$$

Let $D := f(X) - f(\hat{X})$. Then clearly

$$|D| \leq 2a. \tag{8.1.5}$$

Define

$$h(x) := f(x) - f(\hat{x}(x)),$$

so that $D = h(X)$. Note that for $i \neq j$,

$$\frac{\partial \hat{x}_j}{\partial x_i} = \frac{e^{-\Delta_j f(x)}}{(1 + e^{-\Delta_j f(x)})^2} \int_0^1 f_{ij}(x_1, \ldots, x_{j-1}, t, x_{j+1}, \ldots, x_n) \, dt,$$

and for $i = j$, the above derivative is identically equal to zero. Since

$$\frac{e^{-x}}{(1 + e^{-x})^2} \leq \frac{1}{4}$$

for all $x \in \mathbb{R}$, this shows that for all i and j,

$$\left\| \frac{\partial \hat{x}_j}{\partial x_i} \right\| \leq \frac{c_{ij}}{4}. \tag{8.1.6}$$

Thus,

$$\left\| \frac{\partial h}{\partial x_i} \right\| \leq \|f_i\| + \sum_{j=1}^N \|f_j\| \left\| \frac{\partial \hat{x}_j}{\partial x_i} \right\| \tag{8.1.7}$$

$$\leq b_i + \frac{1}{4} \sum_{j=1}^N b_j c_{ij}.$$

Consequently, if $D_i := h(X^{(i)})$, then

$$|D - D_i| \leq b_i + \frac{1}{4} \sum_{j=1}^N b_j c_{ij}. \tag{8.1.8}$$

For $t \in [0, 1]$ and $x \in [0, 1]^N$ define

$$u_i(t, x) := f_i(tx + (1 - t)\hat{x}),$$

so that

$$h(x) = \int_0^1 \sum_{i=1}^N (x_i - \hat{x}_i) u_i(t, x) \, dt.$$

Thus,

$$\mathbb{E}(D^2) = \int_0^1 \sum_{i=1}^N \mathbb{E}((X_i - \hat{X}_i) u_i(t, X) D) \, dt. \tag{8.1.9}$$

Now,

$$\|u_i\| \leq b_i, \tag{8.1.10}$$

and by (8.1.6),

$$\left\|\frac{\partial u_i}{\partial x_i}\right\| \leq t\|f_{ii}\| + (1-t)\sum_{j=1}^{N}\|f_{ij}\|\left\|\frac{\partial \hat{x}_j}{\partial x_i}\right\| \tag{8.1.11}$$

$$\leq tc_{ii} + \frac{1-t}{4}\sum_{j=1}^{N}c_{ij}^2.$$

The bounds (8.1.5), (8.1.8), (8.1.10) and (8.1.11) imply that

$$\left|\mathbb{E}((X_i - \hat{X}_i)u_i(t, X)D) - \mathbb{E}((X_i - \hat{X}_i)u_i(t, X^{(i)})D_i)\right|$$

$$\leq \mathbb{E}\left|\left(u_i(t, X) - u_i(t, X^{(i)})\right)D\right| + \mathbb{E}\left|u_i(t, X^{(i)})(D - D_i)\right|$$

$$\leq 2atc_{ii} + \frac{a(1-t)}{2}\sum_{j=1}^{N}c_{ij}^2 + b_i^2 + \frac{1}{4}\sum_{j=1}^{N}b_ib_jc_{ij}.$$

But $u_i(t, X^{(i)})D_i$ is a function of the random variables $(X_j)_{j\neq i}$ only. Therefore by the definition of $\hat{X}_i$,

$$\mathbb{E}((X_i - \hat{X}_i)u_i(t, X^{(i)})D_i) = 0.$$

Thus,

$$\left|\mathbb{E}((X_i - \hat{X}_i)u_i(t, X)D)\right| \leq 2atc_{ii} + \frac{a(1-t)}{2}\sum_{j=1}^{N}c_{ij}^2 + b_i^2 + \frac{1}{4}\sum_{j=1}^{N}b_ib_jc_{ij}.$$

Using this bound in (8.1.9) gives

$$\mathbb{E}(D^2) \leq \int_0^1 \sum_{i=1}^{N}\left(2atc_{ii} + \frac{a(1-t)}{2}\sum_{j=1}^{N}c_{ij}^2 + b_i^2 + \frac{1}{4}\sum_{j=1}^{N}b_ib_jc_{ij}\right)dt$$

$$= \sum_{i=1}^{N}(ac_{ii} + b_i^2) + \frac{1}{4}\sum_{i,j=1}^{N}(ac_{ij}^2 + b_ib_jc_{ij}),$$

completing the proof. $\qquad\square$

Lemma 8.2 *Let all notation be as in Lemma 8.1. Then*

$$\mathbb{E}\left[\left(\sum_{i=1}^{N}(X_i - \hat{X}_i)\Delta_i f(X)\right)^2\right] \le \sum_{i=1}^{N}b_i^2 + \frac{1}{4}\sum_{i,j=1}^{N}b_i(b_j + 4)c_{ij}.$$

Proof Let g_i denote the function $\Delta_i f$, for notational simplicity. Note that

$$g_i(x) = \int_0^1 f_i(x_1, \dots, x_{i-1}, t, x_{i+1}, \dots, x_N)\, dt,$$

which shows that

$$\|g_i\| \le \|f_i\| = b_i \tag{8.1.12}$$

and for all j,

$$\left\|\frac{\partial g_i}{\partial x_j}\right\| \le \|f_{ij}\| = c_{ij}. \tag{8.1.13}$$

Let

$$G(x) := \sum_{i=1}^{N}(x_i - \hat{x}_i(x))g_i(x).$$

Then

$$\frac{\partial G}{\partial x_i} = \sum_{j=1}^{N}\left[\left(1_{\{j=i\}} - \frac{\partial \hat{x}_j}{\partial x_i}\right)g_j(x) + (x_j - \hat{x}_j)\frac{\partial g_j}{\partial x_i}\right]$$

and therefore by (8.1.6), (8.1.12) and (8.1.13),

$$\left\|\frac{\partial G}{\partial x_i}\right\| \le b_i + \frac{1}{4}\sum_{j=1}^{N}c_{ij}b_j + \sum_{j=1}^{N}c_{ij}. \tag{8.1.14}$$

Note that for any x,

$$|G(x) - G(x^{(i)})| \le \left\|\frac{\partial G}{\partial x_i}\right\|. \tag{8.1.15}$$

Again, $g_i(X)$ and $G(X^{(i)})$ are both functions of $(X_j)_{j\neq i}$ only. Therefore

$$\mathbb{E}((X_i - \hat{X}_i)g_i(X)G(X^{(i)})) = 0. \tag{8.1.16}$$

Combining (8.1.14), (8.1.15) and (8.1.16) gives

$$\mathbb{E}(G(X)^2) = \sum_{i=1}^{N} \mathbb{E}((X_i - \hat{X}_i) g_i(X) G(X))$$

$$\leq \sum_{i=1}^{N} b_i \left(b_i + \frac{1}{4} \sum_{j=1}^{N} c_{ij} b_j + \sum_{j=1}^{N} c_{ij} \right).$$

This completes the proof of the lemma.   □

With the aid of Lemma 8.1 and 8.2, we are now ready to prove Theorem 8.1.

Proof (Proof of the Upper Bound in Theorem 8.1) For $x, y \in [0, 1]^N$, let

$$g(x, y) := \sum_{i=1}^{N} (x_i \log y_i + (1 - x_i) \log(1 - y_i)). \tag{8.1.17}$$

Note that

$$g(x, \hat{x}) - I(\hat{x}) = \sum_{i=1}^{N} (x_i - \hat{x}_i) \log \frac{\hat{x}_i}{1 - \hat{x}_i} = \sum_{i=1}^{N} (x_i - \hat{x}_i) \Delta_i f(x). \tag{8.1.18}$$

Let

$$B := 4 \left(\sum_{i=1}^{N} (ac_{ii} + b_i^2) + \frac{1}{4} \sum_{i,j=1}^{N} (ac_{ij}^2 + b_i b_j c_{ij} + 4b_i c_{ij}) \right)^{1/2}.$$

Let

$$A_1 := \{x \in \{0, 1\}^N : |I(\hat{x}) - g(x, \hat{x})| \leq B/2\},$$

and

$$A_2 := \{x \in \{0, 1\}^N : |f(x) - f(\hat{x})| \leq B/2\}.$$

Let $A = A_1 \cap A_2$. By Lemma 8.1 and the identity (8.1.18), $\mathbb{P}(X \notin A_1) \leq 1/4$. By Lemma 8.2, $\mathbb{P}(X \notin A_2) \leq 1/4$. Thus, $\mathbb{P}(X \in A) \geq 1/2$, which is the same as

$$\frac{\sum_{x \in A} e^{f(x)}}{\sum_{x \in \{0,1\}^N} e^{f(x)}} \geq \frac{1}{2}.$$

Therefore by the definition of the set A,

$$F = \log \sum_{x \in \{0,1\}^N} e^{f(x)} \leq \log \sum_{x \in A} e^{f(x)} + \log 2 \qquad (8.1.19)$$

$$\leq B + \log \sum_{x \in A} e^{f(\hat{x}) - I(\hat{x}) + g(x,\hat{x})} + \log 2.$$

Now take some $x \in \{0,1\}^N$ and let d satisfy (8.1.1). Then by the Cauchy–Schwarz inequality,

$$\sum_{i=1}^{N} |f_i(x) - d_i| \leq N\epsilon.$$

Fix such an x and d. Note that for each i,

$$|\Delta_i f(x) - f_i(x)| \leq \int_0^1 |f_i(x_1, \ldots, x_{i-1}, t, x_{i+1}, \ldots, x_N) - f_i(x)| \, dt$$

$$\leq \|f_{ii}\| = c_{ii}.$$

By the last two inequalities and (8.1.1),

$$\sum_{i=1}^{N} |\Delta_i f(x) - d_i| \leq N\epsilon + \sum_{i=1}^{N} c_{ii}. \qquad (8.1.20)$$

and

$$\left(\sum_{i=1}^{N} (\Delta_i f(x) - d_i)^2 \right)^{1/2} \leq N^{1/2} \epsilon + \left(\sum_{i=1}^{N} c_{ii}^2 \right)^{1/2}. \qquad (8.1.21)$$

Let $u(x) = 1/(1 + e^{-x})$. Note that for all x,

$$|u'(x)| = \frac{1}{(e^{x/2} + e^{-x/2})^2} \leq \frac{1}{4}.$$

Therefore if a vector $p = p(d)$ is defined as $p_i = u(d_i)$, then by (8.1.21),

$$\left(\sum_{i=1}^{N} (\hat{x}_i - p_i)^2 \right)^{1/2} \leq \left(\frac{1}{16} \sum_{i=1}^{N} (\Delta_i f(x) - d_i)^2 \right)^{1/2}$$

$$\leq \frac{N^{1/2} \epsilon}{4} + \frac{1}{4} \left(\sum_{i=1}^{N} c_{ii}^2 \right)^{1/2}.$$

Thus, if

$$L := \Big(\sum_{i=1}^{N} b_i^2 \Big)^{1/2},$$

then

$$|f(\hat{x}) - f(p)| \le L \Big(\sum_{i=1}^{N} (\hat{x}_i - p_i)^2 \Big)^{1/2} \tag{8.1.22}$$

$$\le \frac{L N^{1/2} \epsilon}{4} + \frac{L}{4} \Big(\sum_{i=1}^{N} c_{ii}^2 \Big)^{1/2}.$$

Next, let $v(x) = \log(1 + e^{-x})$. Then for all x,

$$|v'(x)| = \frac{e^{-x}}{1 + e^{-x}} \le 1.$$

Consequently,

$$|\log \hat{x}_i - \log p_i| \le |\Delta_i f(x) - d_i|$$

and

$$|\log(1 - \hat{x}_i) - \log(1 - p_i)| \le |\Delta_i f(x) - d_i|.$$

Therefore by (8.1.20),

$$|g(x, \hat{x}) - g(x, p)| \le 2 \sum_{i=1}^{N} |\Delta_i f(x) - d_i| \le 2N\epsilon + 2 \sum_{i=1}^{N} c_{ii}. \tag{8.1.23}$$

Finally, let $w(x) = I(u(x))$. Then

$$w'(x) = u'(x) I'(u(x))$$

$$= \frac{e^{-x}}{(1 + e^{-x})^2} \log \frac{u(x)}{1 - u(x)}$$

$$= \frac{x e^{-x}}{(1 + e^{-x})^2}.$$

Thus, for all x,

$$|w'(x)| \le \sup_{x \in \mathbb{R}} \frac{|x| e^{-x}}{(1 + e^{-x})^2} \le \sup_{x \ge 0} x e^{-x} = \frac{1}{e}.$$

Consequently,

$$|I(\hat{x}_i) - I(p_i)| \le \frac{1}{e}|\Delta_i f(x) - d_i|,$$

and so by (8.1.20),

$$|I(\hat{x}) - I(p)| \le \frac{N\epsilon}{e} + \frac{1}{e}\sum_{i=1}^{N} c_{ii}. \qquad (8.1.24)$$

For each $d \in \mathscr{D}(\epsilon)$ let $\mathscr{C}(d)$ be the set of all $x \in \{0, 1\}^N$ such that (8.1.1) holds, and let $p(d)$ be the vector p defined above. Then by (8.1.22), (8.1.23) and (8.1.24),

$$\log \sum_{x \in A} e^{f(\hat{x}) - I(\hat{x}) + g(x, \hat{x})} \le \log \sum_{d \in \mathscr{D}(\epsilon)} \sum_{x \in \mathscr{C}(d)} e^{f(\hat{x}) - I(\hat{x}) + g(x, \hat{x})} \qquad (8.1.25)$$

$$\le \frac{LN^{1/2}\epsilon}{4} + \frac{L}{4}\Big(\sum_{i=1}^{N} c_{ii}^2\Big)^{1/2} + 2N\epsilon + 2\sum_{i=1}^{N} c_{ii} + \frac{N\epsilon}{e} + \frac{1}{e}\sum_{i=1}^{N} c_{ii}$$

$$+ \log \sum_{d \in \mathscr{D}(\epsilon)} \sum_{x \in \mathscr{C}(d)} e^{f(p(d)) - I(p(d)) + g(x, p(d))}.$$

Now note that for any $p \in [0, 1]^N$,

$$\sum_{x \in \{0,1\}^N} e^{g(x,p)} = 1.$$

Thus,

$$\log \sum_{d \in \mathscr{D}(\epsilon)} \sum_{x \in \mathscr{C}(d)} e^{f(p(d)) - I(p(d)) + g(x, p(d))} \qquad (8.1.26)$$

$$\le \log \sum_{d \in \mathscr{D}(\epsilon)} e^{f(p(d)) - I(p(d))}$$

$$\le \log |\mathscr{D}(\epsilon)| + \sup_{p \in [0,1]^N} (f(p) - I(p)).$$

Combining (8.1.19), (8.1.25) and (8.1.26), the proof is complete.   □

Proof (Proof of the Lower Bound in Theorem 8.1) Let y be a point in the cube $[0, 1]^N$. Let $Y = (Y_1, \ldots, Y_N)$ be a random vector with independent components, where Y_i is a Bernoulli(y_i) random variable. Then by Jensen's inequality,

$$\sum_{x\in\{0,1\}^N} e^{f(x)} = \sum_{x\in\{0,1\}^N} e^{f(x)-g(x,y)+g(x,y)}$$

$$= \mathbb{E}(e^{f(Y)-g(Y,y)})$$

$$\geq \exp(\mathbb{E}(f(Y) - g(Y,y)))$$

$$= \exp(\mathbb{E}(f(Y)) - I(y)).$$

Let $S := f(Y) - f(y)$. For $t \in [0, 1]$ and $x \in [0, 1]^N$ define

$$v_i(t, x) := f_i(tx + (1 - t)y),$$

so that

$$S = \int_0^1 \sum_{i=1}^N (Y_i - y_i) v_i(t, Y)\, dt. \tag{8.1.27}$$

By the independence of Y_i and $Y^{(i)}$,

$$\left|\mathbb{E}((Y_i - y_i) v_i(t, Y))\right| = \left|\mathbb{E}((Y_i - y_i)(v_i(t, Y) - v_i(t, Y^{(i)})))\right|$$

$$\leq \left\|\frac{\partial v_i}{\partial x_i}\right\| \leq t c_{ii}.$$

Using this bound in (8.1.27) gives

$$\mathbb{E}(S) \geq -\int_0^1 \sum_{i=1}^N t c_{ii}\, dt = -\frac{1}{2} \sum_{i=1}^N c_{ii}.$$

This completes the proof.   □

8.2 Gradient Complexity of Homomorphism Densities

The goal of this section is to compute estimates for the smoothness and complexity terms of Theorem 8.1 for homomorphism densities. These estimates will be used later for establishing quantitative versions of some of the theorems proved earlier in this monograph, which, in turn, will be used for understanding large deviations in sparse graphs.

Let $n \geq 2$ be a positive integer. Let $\mathscr{P}_n$ denote the set of all upper triangular arrays like $x = (x_{ij})_{1\leq i<j\leq n}$, where each $x_{ij} \in [0, 1]$. Let H be a finite simple graph with vertex set $V = \{1, 2, \dots, k\}$ and edge set E. For $x \in \mathscr{P}_n$, let

$$t(H, x) := \frac{1}{n^k} \sum_{q_1,\dots,q_k=1}^n \prod_{\{l,l'\}\in E} x_{q_l q_{l'}}, \tag{8.2.1}$$

where x_{ii} is interpreted as zero for each i and x_{ij} denotes x_{ji} if $i > j$. Let

$$f(x) := n^2 t(H, x). \tag{8.2.2}$$

We will now obtain estimates for the smoothness and complexity terms for this f. The smoothness term is straightforward. The estimates for the smoothness term are given in Theorem 8.2 below. The bulk of this section is devoted to the analysis of the complexity term. The estimates for the complexity term are given in Theorem 8.3.

Theorem 8.2 *Let H and f be as above, and let $m := |E|$. Then $\|f\| \leq n^2$, and for any $i < j$ and $i' < j'$,*

$$\left\| \frac{\partial f}{\partial x_{ij}} \right\| \leq 2m, \text{ and}$$

$$\left\| \frac{\partial^2 f}{\partial x_{ij} \partial x_{i'j'}} \right\| \leq \begin{cases} 4m(m-1)n^{-1} & \text{if } |\{i,j,i',j'\}| = 2 \text{ or } 3, \\ 4m(m-1)n^{-2} & \text{if } |\{i,j,i',j'\}| = 4. \end{cases}$$

Proof It is clear that $\|f\| \leq n^2$ since the x_{ij}'s are all in $[0,1]$ and there are exactly n^k terms in the sum that defines f. Next, note that for any $i < j$,

$$\frac{\partial f}{\partial x_{ij}} = \frac{1}{n^{k-2}} \sum_{\substack{\{a,b\} \in E}} \sum_{\substack{q \in [n]^k \\ \{q_a,q_b\}=\{i,j\}}} \prod_{\substack{\{l,l'\} \in E \\ \{l,l'\} \neq \{a,b\}}} x_{q_l q_{l'}}, \tag{8.2.3}$$

and therefore

$$\left\| \frac{\partial f}{\partial x_{ij}} \right\| \leq \frac{2mn^{k-2}}{n^{k-2}} = 2m.$$

Next, for any $i < j$ and $i' < j'$,

$$\frac{\partial^2 f}{\partial x_{ij} \partial x_{i'j'}} = \frac{1}{n^{k-2}} \sum_{\substack{\{a,b\} \in E}} \sum_{\substack{\{c,d\} \in E \\ \{c,d\} \neq \{a,b\}}} \sum_{\substack{q \in [n]^k \\ \{q_a,q_b\}=\{i,j\} \\ \{q_c,q_d\}=\{i',j'\}}} \prod_{\substack{\{l,l'\} \in E \\ \{l,l'\} \neq \{a,b\} \\ \{l,l'\} \neq \{c,d\}}} x_{q_l q_{l'}}.$$

Take any two edges $\{a,b\}, \{c,d\} \in E$ such that $\{a,b\} \neq \{c,d\}$. Then the number of choices of $q \in [n]^k$ such that $\{q_a, q_b\} = \{i,j\}$ and $\{q_c, q_d\} = \{i',j'\}$ is at most $4n^{k-3}$ if $|\{i,j,i',j'\}| = 2$ or 3 (since we are constraining q_a, q_b, q_c and q_d and $|\{a,b,c,d\}| \geq 3$ always), and at most $4n^{k-4}$ if $|\{i,j,i',j'\}| = 4$ (since $|\{a,b,c,d\}|$ must be 4 if there is at least one possible choice of q for these i,j,i',j'). This gives the upper bound for the second derivatives. $\square$

To understand the complexity of the gradient of f, we need some preparation. For an $n \times n$ matrix M, recall the definition of the operator norm:

$$\|M\| := \max\{\|Mz\| : z \in \mathbb{R}^n, \|z\| = 1\},$$

where $\|z\|$ denotes the Euclidean norm of z. Equivalently, $\|M\|$ is the largest singular value of M. For $x \in \mathscr{P}_n$, let $M(x)$ be the symmetric matrix whose (i,j)th entry is x_{ij}, with the convention that $x_{ij} = x_{ji}$ and $x_{ii} = 0$. Define the operator norm of x as

$$\|x\|_{\mathrm{op}} := \|M(x)\|.$$

The following lemma estimates the entropy of the unit cube under this norm.

Lemma 8.3 *For any $\tau \in (0, 1)$, there is a finite set of $n \times n$ matrices $\mathscr{W}(\tau)$ such that*

$$|\mathscr{W}(\tau)| \le e^{34(n/\tau^2)\log(51/\tau^2)},$$

and for any $n \times n$ matrix M with entries in $[0, 1]$, there exists $W \in \mathscr{W}(\tau)$ such that

$$\|M - W\| \le n\tau.$$

In particular, for any $x \in \mathscr{P}_n$ there exists $W \in \mathscr{W}(\tau)$ such that

$$\|M(x) - W\| \le n\tau.$$

Proof Suppose that $n < 17/\tau^2$. Let $\mathscr{W}(\tau)$ consist of all $n \times n$ matrices whose entries are all integer multiples of τ and belong to the interval $[0, 1]$. It is then easy to see that for any M with entries in $[0, 1]$, there exists $W \in \mathscr{W}(\tau)$ such that $\|M - W\| \le n\tau$. Moreover, since $n < 17/\tau^2$,

$$|\mathscr{W}(\tau)| \le \tau^{-n^2} \le e^{9(n/\tau^2)\log(1/\tau^2)},$$

completing the proof in this case.

Next, suppose that $n \ge 17/\tau^2$. Let l be the integer part of $17/\tau^2$ and $\delta = 1/l$. Let $\mathscr{A}$ be a finite subset of the unit ball of $\mathbb{R}^n$ such that any vector inside the ball is at Euclidean distance $\le \delta$ from some element of $\mathscr{A}$. (In other words, $\mathscr{A}$ is a δ-net of the unit ball under the Euclidean metric.) The set $\mathscr{A}$ may be defined as a maximal set of points in the unit ball such that any two are at a distance greater than δ from each other. Since the balls of radius $\delta/2$ around these points are disjoint and their union is contained in the ball of radius $1 + \delta/2$ centered at zero, it follows that $|\mathscr{A}|C(\delta/2)^n \le C(1 + \delta/2)^n$, where C is the volume of the unit ball. Therefore,

$$|\mathscr{A}| \le (3/\delta)^n. \tag{8.2.4}$$

Take any $x \in \mathscr{P}_n$. Suppose that M has singular value decomposition

$$M = \sum_{i=1}^{n} \lambda_i u_i v_i^t,$$

where $\lambda_1 \geq \lambda_2 \geq \cdots \lambda_n \geq 0$ are the singular values of M, and $u_1, \ldots, u_n$ and $v_1, \ldots, v_n$ are singular vectors, and v_i^t denotes the transpose of the column vector v_i. Assume that the u_i's and v_i's are orthonormal systems. Since the elements of M all belong to the interval $[0, 1]$, it is easy to see that $\lambda_1 \leq n$ and $\sum \lambda_i^2 \leq n^2$. Due to the second inequality, there exists $y \in \mathscr{A}$ such that

$$\sum_{i=1}^{n} (n^{-1}\lambda_i - y_i)^2 \leq \delta^2. \tag{8.2.5}$$

Let $z_1, \ldots, z_n$ and $w_1, \ldots, w_n$ be elements of $\mathscr{A}$ such that for each i,

$$\sum_{j=1}^{n} (u_{ij} - z_{ij})^2 \leq \delta^2 \quad \text{and} \quad \sum_{j=1}^{n} (v_{ij} - w_{ij})^2 \leq \delta^2, \tag{8.2.6}$$

where u_{ij} denotes the jth component of the vector u_i, etc. Define two matrices V and W as

$$V := \sum_{i=1}^{l-1} \lambda_i u_i v_i^t \quad \text{and} \quad W := \sum_{i=1}^{l-1} n y_i z_i w_i^t.$$

Note that since $\sum \lambda_i^2 \leq n^2$ and λ_i decreases with i, therefore for each i, $\lambda_i^2 \leq n^2/i$. Thus,

$$\|M - W\| \leq \|M - V\| + \|V - W\|$$

$$\leq \frac{n}{\sqrt{l}} + \|V - W\|.$$

Next, note that by (8.2.6), the operator norms of the rank-one matrices $(u_i - z_i)v_i^t$ and $z_i(v_i - w_i)^t$ are bounded by δ. And by (8.2.5), $|\lambda_i - n y_i| \leq n\delta$ for each i. Therefore

$$\|V - W\| \leq \left\| \sum_{i=1}^{l-1} (\lambda_i - n y_i) u_i v_i^t \right\| + \left\| \sum_{i=1}^{l-1} n y_i (u_i - z_i) v_i^t \right\|$$

$$+ \left\| \sum_{i=1}^{l-1} n y_i z_i (v_i - w_i)^t \right\|$$

$$\leq \max_{1 \leq i \leq l-1} |\lambda_i - n y_i| + 2 \sum_{i=1}^{l-1} n |y_i| \delta$$

$$\leq n\delta + 2n\delta \left((l-1) \sum_{i=1}^{l-1} y_i^2 \right)^{1/2} \leq n\delta + 2n\delta\sqrt{l-1} \leq \frac{3n}{\sqrt{l}}.$$

Thus,

$$\|M - W\| \le \frac{4n}{\sqrt{l}} \le \frac{4n}{\sqrt{\frac{17}{\tau^2} - 1}} \le \frac{4n}{\sqrt{\frac{16}{\tau^2}}} = n\tau.$$

Let $\mathscr{W}(\tau)$ be the set of all possible W's constructed in the above manner. Then $\mathscr{W}(\tau)$ has the required property, and by (8.2.4),

$$|\mathscr{W}(\tau)| \le \text{The number of ways of choosing}$$

$$y, z_1, \ldots, z_{l-1}, w_1, \ldots, w_{l-1} \in \mathscr{A}$$

$$= |\mathscr{A}|^{2l-1} \le (3/\delta)^{2nl} = e^{2nl \log(3l)}.$$

This completes the proof of the lemma. $\qquad\qquad\qquad\qquad\qquad\qquad\square$

Let r be a positive integer. Let K_r be the complete graph on the vertex set $\{1, \ldots, r\}$. For any set of edges A of K_r, any $q = (q_1, \ldots, q_r) \in [n]^r$, and any $x \in \mathscr{P}_n$, let

$$P(x, q, A) := \prod_{\{a,b\} \in A} x_{q_a q_b},$$

with the usual convention that the empty product is 1. Note that if $q_a = q_b$ for some $\{a, b\} \in A$, the $P(x, q, A) = 0$ due to our convention that $x_{ii} = 0$ for each i. Next, note that if A and B are disjoint sets of edges, then

$$P(x, q, A \cup B) = P(x, q, A)P(x, q, B). \tag{8.2.7}$$

Lemma 8.4 *Let A and B be sets of edges of K_r, and let $e = \{\alpha, \beta\}$ be an edge that is neither in A nor in B. Then for any $x, y \in \mathscr{P}_n$,*

$$\left| \sum_{q \in [n]^r} P(x, q, A)P(y, q, B)(x_{q_\alpha q_\beta} - y_{q_\alpha q_\beta}) \right| \le n^{r-1} \|x - y\|_{\mathrm{op}}.$$

Proof By relabeling the vertices of K_r and redefining A and B, we may assume that $\alpha = 1$ and $\beta = 2$.

Let A_1 be the set of all edges in A that are incident to 1. Let A_2 be the set of all edges in A that are incident to 2. Note that since $\{1, 2\} \notin A$, therefore A_1 and A_2 must be disjoint. Similarly, let B_1 be the set of all edges in B that are incident to 1 and let B_2 be the set of all edges in B that are incident to 2. Let $A_3 = A \setminus (A_1 \cup A_2)$ and $B_3 = B \setminus (B_1 \cup B_2)$. By (8.2.7),

$$P(x, q, A) = P(x, q, A_1)P(x, q, A_2)P(x, q, A_3)$$

and

$$P(y, q, B) = P(y, q, B_1)P(y, q, B_2)P(y, q, B_3).$$

Thus,

$$\sum_{q \in [n]^r} P(x, q, A)P(y, q, B)(x_{q_1 q_2} - y_{q_1 q_2})$$

$$= \sum_{q_3, \dots, q_r} P(x, q, A_3)P(y, q, B_3)\left(\sum_{q_1, q_2} Q(x, y, q)(x_{q_1 q_2} - y_{q_1 q_2}) \right),$$

where

$$Q(x, y, q) = P(x, q, A_1)P(x, q, A_2)P(y, q, B_1)P(y, q, B_2).$$

Now fix $q_3, \dots, q_r$. Then $P(x, q, A_1)P(y, q, B_1)$ is a function of q_1 only, and does not depend on q_2. Let $g(q_1)$ denote this function. Similarly, $P(x, q, A_2)P(y, q, B_2)$ is a function of q_2 only, and does not depend on q_1. Let $h(q_2)$ denote this function. Both g and h are uniformly bounded by 1. Therefore

$$\left| \sum_{q_1, q_2} Q(x, y, q)(x_{q_1 q_2} - y_{q_1 q_2}) \right| = \left| \sum_{q_1, q_2} g(q_1)h(q_2)(x_{q_1 q_2} - y_{q_1 q_2}) \right|$$

$$\leq n\|x - y\|_{\mathrm{op}}.$$

Since this is true for all choices of $q_3, \dots, q_r$ and P is also uniformly bounded by 1, this completes the proof of the lemma.   □

Let A and B be two sets of edges of K_r. For $x, y \in \mathscr{P}_n$, define

$$R(x, y, A, B) := \sum_{q \in [n]^r} P(x, q, A)P(y, q, B).$$

Lemma 8.5 *Let A, B, A' and B' be sets of edges of K_r such that $A \cap B = A' \cap B' = \emptyset$ and $A \cup B = A' \cup B'$. Then*

$$|R(x, y, A, B) - R(x, y, A', B')| \leq \frac{1}{2}r(r - 1)n^{r-1}\|x - y\|_{\mathrm{op}}.$$

Proof First, suppose that $e = \{\alpha, \beta\}$ is an edge such that $e \notin A'$ and $A = A' \cup \{e\}$. Since $A \cup B = A' \cup B'$ and $A \cap B = A' \cap B' = \emptyset$, this implies that $e \notin B$ and $B' = B \cup \{e\}$. Thus,

$$R(x, y, A, B) - R(x, y, A', B') = \sum_{q \in [n]^r} P(x, q, A')P(y, q, B)(x_{q_\alpha q_\beta} - y_{q_\alpha q_\beta}),$$

and the proof is completed using Lemma 8.4. For the general case, simply 'move' from the pair (A, B) to the pair (A', B') by 'moving one edge at a time' and apply Lemma 8.4 at each step.   □

Lemma 8.6 *Let $g_{ij} := \partial f / \partial x_{ij}$, where f is the function defined in Eq. (8.2.2), and let $m := |E|$, as before. Then for any $x, y \in \mathscr{P}_n$,*

$$\sum_{1 \leq i < j \leq n} (g_{ij}(x) - g_{ij}(y))^2 \leq 8m^2 k^2 n \|x - y\|_{\mathrm{op}}.$$

Proof Recall Eq. (8.2.3), that is, for any $1 \leq i < j \leq N$,

$$g_{ij}(x) = \frac{\partial f}{\partial x_{ij}} = \frac{1}{n^{k-2}} \sum_{\{a,b\} \in E} \sum_{\substack{q \in [n]^k \\ \{q_a, q_b\} = \{i,j\}}} \prod_{\substack{\{l,l'\} \in E \\ \{l,l'\} \neq \{a,b\}}} x_{q_l q_{l'}}.$$

Although differentiating with respect to x_{ii} does not make sense, let g_{ii} be the function defined using the same formula as above. When $i > j$, let $g_{ij} = g_{ji}$. Fix $x, y \in \mathscr{P}_n$. Define for any $q \in [n]^k$ and $\{a, b\} \in E$

$$D(q, \{a, b\}) := \prod_{\substack{\{l,l'\} \in E \\ \{l,l'\} \neq \{a,b\}}} x_{q_l q_{l'}} - \prod_{\substack{\{l,l'\} \in E \\ \{l,l'\} \neq \{a,b\}}} y_{q_l q_{l'}}.$$

Define

$$\theta_{ij} := \begin{cases} 2 & \text{if } i = j, \\ 1/2 & \text{if } i \neq j, \end{cases} \qquad \gamma_{ij} := \begin{cases} 2 & \text{if } i = j, \\ 1 & \text{if } i \neq j. \end{cases}$$

Then note that

$$\sum_{i,j=1}^{n} \theta_{ij} (g_{ij}(x) - g_{ij}(y))^2$$

$$= \frac{1}{n^{2k-4}} \sum_{i,j=1}^{n} \theta_{ij} \left(\sum_{\{a,b\} \in E} \sum_{\substack{q \in [n]^k \\ \{q_a, q_b\} = \{i,j\}}} D(q, \{a, b\}) \right)^2$$

$$= \frac{1}{n^{2k-4}} \sum_{i,j=1}^{n} \sum_{\substack{\{a,b\} \in E \\ \{c,d\} \in E}} \sum_{\substack{q \in [n]^k \\ \{q_a, q_b\} = \{i,j\}}} \sum_{\substack{s \in [n]^k \\ \{s_c, s_d\} = \{i,j\}}} \theta_{ij} D(q, \{a, b\}) D(s, \{c, d\})$$

$$= \frac{1}{n^{2k-4}} \sum_{\substack{\{a,b\} \in E \\ \{c,d\} \in E}} \sum_{q \in [n]^k} \sum_{\substack{s \in [n]^k \\ \{s_c, s_d\} = \{q_a, q_b\}}} \gamma_{q_a q_b} D(q, \{a, b\}) D(s, \{c, d\}).$$

Now fix two edges $\{a, b\}$ and $\{c, d\}$ in E. Relabeling vertices if necessary, assume that $c = k - 1$ and $d = k$. Let $r = 2k - 2$. For any $t \in [n]^r$, define two vectors $q(t)$ and $s(t)$ in $[n]^k$ as follows. For $i = 1, \ldots, k$, let $q_i(t) = t_i$. For $i = 1, \ldots, k - 2$, let $s_i(t) = t_{i+k}$. Let $s_{k-1}(t) = t_a$ and $s_k(t) = t_b$. Note that

$$\sum_{q \in [n]^k} \sum_{\substack{s \in [n]^k \\ \{s_c, s_d\} = \{q_a, q_b\}}} \gamma_{q_a q_b} D(q, \{a, b\}) D(s, \{c, d\})$$

$$= \sum_{q \in [n]^k} \sum_{\substack{s \in [n]^k \\ s_c = q_a, s_d = q_b}} D(q, \{a, b\}) D(s, \{c, d\})$$

$$+ \sum_{q \in [n]^k} \sum_{\substack{s \in [n]^k \\ s_c = q_b, s_d = q_a}} D(q, \{a, b\}) D(s, \{c, d\}).$$

With the definition of $q(t)$ and $s(t)$ given above, it is clear that the first term on the right-hand side equals

$$\sum_{t \in [n]^r} D(q(t), \{a, b\}) D(s(t), \{c, d\}).$$

Below, we will get a bound on this term. The same upper bound will hold for the other term by symmetry.

Define two subsets of edges A and B of K_r as follows. Let A be the set of all edges $\{l, l'\}$ such that $\{l, l'\} \in E \setminus \{\{a, b\}\}$. Let B be the set of all edges $\{\phi(l), \phi(l')\}$ such that $\{l, l'\} \in E \setminus \{\{k - 1, k\}\}$, where $\phi : [k] \to [r]$ is the map

$$\phi(x) = \begin{cases} x + k & \text{if } x \neq k - 1 \text{ and } x \neq k, \\ a & \text{if } x = k - 1, \\ b & \text{if } x = k. \end{cases}$$

By the above construction, $q_l(t) = t_l$ and $s_l(t) = t_{\phi(l)}$. Therefore it is easy to see, for instance, that

$$\sum_{t \in [n]^r} \prod_{\substack{\{l, l'\} \in E \\ \{l, l'\} \neq \{a, b\}}} x_{q_l(t) q_{l'}(t)} \prod_{\substack{\{l, l'\} \in E \\ \{l, l'\} \neq \{k-1, k\}}} y_{s_l(t) s_{l'}(t)}$$

$$= \sum_{t \in [n]^r} \prod_{\substack{\{l, l'\} \in E \\ \{l, l'\} \neq \{a, b\}}} x_{t_l t_{l'}} \prod_{\substack{\{l, l'\} \in E \\ \{l, l'\} \neq \{k-1, k\}}} y_{t_{\phi(l)} t_{\phi(l')}}$$

$$= R(x, y, A, B).$$

Carrying out similar computations for the other terms, we get

$$\sum_{t \in [n]^r} D(q(t), \{a, b\}) D(s(t), \{c, d\})$$

$$= R(x, y, A \cup B, \emptyset) - R(x, y, B, A) - R(x, y, A, B) + R(x, y, \emptyset, A \cup B).$$

Lastly, note that $A \cap B = \emptyset$ since for any $\{l, l'\} \in E \setminus \{\{k-1, k\}\}$, at least one among $\phi(l)$ and $\phi(l')$ must be strictly bigger than k and therefore $\{\phi(l), \phi(l')\}$ cannot be an element of A. The proof is now completed by applying Lemma 8.5. $\square$

With the help of Lemma 8.3 and Lemma 8.6, we are now ready to prove the following theorem, which gives an estimate for the complexity of the gradient of f.

Theorem 8.3 *Let $m := |E|$, as before. For the function f defined in Eq. (8.2.2), one can produce sets $\mathscr{D}(\epsilon)$ satisfying the criterion (8.1.1) (with $N = n(n-1)/2$) such that*

$$|\mathscr{D}(\epsilon)| \leq \exp\left(\frac{C_1 m^4 k^4 n}{\epsilon^4} \log \frac{C_2 m^4 k^4}{\epsilon^4}\right),$$

where C_1 and C_2 are universal constants.

Proof Take any $\epsilon > 0$ and let

$$\tau = \frac{\epsilon^2}{64 m^2 k^2}.$$

Let $\mathscr{W}(\tau)$ be as in Lemma 8.3. For each $W \in \mathscr{W}(\tau)$, let $y(W) \in \mathscr{P}_n$ be a vector such that $\|M(y) - W\| \leq n\tau$. If for some W there does not exist any such y, leave $y(W)$ undefined. Let $g_{ij} = \partial f / \partial x_{ij}$, as in Lemma 8.6. Let $g : \mathscr{P}_n \to \mathbb{R}^{n(n-1)/2}$ be the function whose (i, j)th coordinate is g_{ij}. Define

$$\mathscr{D}(\epsilon) := \{g(y) : y = y(W) \text{ for some } W \in \mathscr{W}(\tau)\}.$$

Then by Lemma 8.3

$$|\mathscr{D}(\epsilon)| \leq |\mathscr{W}(\tau)| \leq e^{34(n/\tau^2) \log(51/\tau^2)}.$$

We claim that the set $\mathscr{D}(\epsilon)$ satisfies the requirements of Theorem 8.5. To see this, take any $x \in \mathscr{P}_n$. By Lemma 8.3, there exists $W \in \mathscr{W}(\tau)$ such that $\|M(x) - W\| \leq n\tau$. In particular, this means that $y := y(W)$ is defined, and so

$$\|x - y\|_{\text{op}} = \|M(x) - M(y)\|$$

$$\leq \|M(x) - W\| + \|W - M(y)\|$$

$$\leq 2n\tau.$$

Therefore by Lemma 8.6,

$$\sum_{1 \le i < j \le n} (g_{ij}(x) - g_{ij}(y))^2 \le 16m^2k^2n^2\tau.$$

Let $z = g(x)$ and $v = g(y)$. Then $v \in \mathcal{D}(\epsilon)$, and by the above inequality,

$$\sum_{1 \le i < j \le n} (z_{ij} - v_{ij})^2 \le 16m^2k^2n^2\tau = \frac{n^2\epsilon^2}{4} \le \binom{n}{2}\epsilon^2.$$

This proves the claim that $\mathcal{D}(\epsilon)$ satisfies the requirements of Theorem 8.5. This completes the proof of Theorem 8.3.   □

8.3   Quantitative Estimates for Exponential Random Graphs

Take a positive integer k. Let $H_1, \ldots, H_k$ be finite simple graphs and let $\beta_1, \ldots, \beta_k$ be real numbers. Consider the exponential random graph on n vertices defined by the statistic T from Eq. (7.4.1) of Chap. 7. Let ψ_n be the normalizing constant of this model, as defined in Eq. (7.1.1) of Chap. 7. Let $\mathcal{P}_n$ be as in the previous section. Recall the function I defined in Eq. (8.1.2). For $x \in \mathcal{P}_n$, let

$$I(x) := \sum_{1 \le i < j \le n} I(x_{ij}).$$

Let

$$L_n := \sup_{x \in \mathcal{P}_n} \left(\beta_1 t(H_1, x) + \cdots + \beta_k t(H_k, x) - \frac{I(x)}{n^2} \right).$$

The following theorem gives a quantitative version of Theorem 7.1 of Chap. 7 for this kind of T. This allows the possibility of extending the analysis of exponential random graphs to the sparse regime. The proof is based on an application of Theorem 8.1 using the estimates from Theorem 8.2 and Theorem 8.3.

Theorem 8.4 *Let ψ_n and L_n be as above. Let $B := 1 + |\beta_1| + \cdots + |\beta_k|$. Then*

$$-\frac{cB}{n} \le \psi_n - L_n \le \frac{CB^{8/5}(\log n)^{1/5}}{n^{1/5}}\left(1 + \frac{\log B}{\log n}\right) + \frac{CB^2}{n^{1/2}},$$

where C and c are constants that depend only on $H_1, \ldots, H_k$.

Proof Throughout this proof, C will denote any constant that depends only on the graphs $H_1, \ldots, H_k$. As in the previous section, $N = n(n-1)/2$. For $1 \leq j \leq k$, define

$$T_j(x) := n^2 t(H_j, x),$$

where $t(H_j, x)$ is defined as in Eq. (8.2.1). Let

$$f(x) := \beta_1 T_1(x) + \cdots + \beta_l T_k(x),$$

so that $\psi_n = F$ for this f, in the notation of Theorem 8.1. Let a, $b_{(ij)}$ and $c_{(ij)(i'j')}$ be as in Theorem 8.1, with obvious notational modification. Then

$$a \leq n^2 \sum_{r=1}^{k} |\beta_r| \leq CBn^2.$$

By Theorem 8.2, we get the estimates

$$b_{(ij)} \leq CB$$

and

$$c_{(ij)(i'j')} \leq \begin{cases} CBn^{-1} & \text{if } |\{i, j, i', j'\}| = 2 \text{ or } 3, \\ CBn^{-2} & \text{if } |\{i, j, i', j'\}| = 4. \end{cases}$$

Let $\mathscr{D}_1(\epsilon), \ldots, \mathscr{D}_k(\epsilon)$ be the $\mathscr{D}(\epsilon)$'s for $T_1, \ldots, T_k$. Define

$$\mathscr{D}(\epsilon) := \{\beta_1 d_1 + \cdots + \beta_k d_k : d_r \in \mathscr{D}_i(\epsilon/\beta_r k), \ r = 1, \ldots, k\}.$$

Clearly, for any $x \in \mathscr{P}_n$, there exists $d_1 \in \mathscr{D}_1(\epsilon/\beta_1 k), \ldots, d_k \in \mathscr{D}_k(\epsilon/\beta_k k)$ such that (again, with obvious notation)

$$\sum_{(ij)} (f_{(ij)}(x) - (\beta_1 d_{1(ij)} + \cdots + \beta_k d_{k(ij)}))^2$$

$$\leq k \sum_{r=1}^{k} \sum_{(ij)} \beta_r^2 (T_{r(ij)}(x) - d_{r(ij)})^2 \leq N\epsilon^2.$$

Therefore, $\mathscr{D}(\epsilon)$ satisfies the requirement of Theorem 8.1. Also,

$$|\mathscr{D}(\epsilon)| \leq \prod_{r=1}^{k} |\mathscr{D}_r(\epsilon/\beta_r k)|. \tag{8.3.1}$$

By the bounds on a, $b_{(ij)}$ and $c_{(ij)(i'j')}$ obtained in the previous section, the following estimates are easy:

$$\sum_{(ij)} ac_{(ij)(ij)} \leq CB^2 n^3, \quad \sum_{(ij)} b_{(ij)}^2 \leq CB^2 n^2,$$

$$\sum_{(ij),(i'j')} ac_{(ij)(i'j')}^2 \leq CB^3 n^3,$$

$$\sum_{(ij),(i'j')} b_{(ij)}(b_{(i'j')} + 4)c_{(ij)(i'j')} \leq CB^3 n^2,$$

$$\sum_{(ij)} c_{(ij)(ij)}^2 \leq CB^2, \quad \sum_{(ij)} c_{(ij)(ij)} \leq CBn.$$

Combining these estimates, we see that the smoothness term is bounded by $CB^2 n^{3/2}$. Next, by (8.3.1) and Theorem 8.3,

$$\log |\mathscr{D}(\epsilon)| \leq \sum_{r=1}^{k} \log |\mathscr{D}_r(\epsilon/\beta_r k)|$$

$$\leq \frac{CB^4 n}{\epsilon^4} \log \frac{CB^4}{\epsilon^4}.$$

Therefore, the complexity term (of Theorem 8.1) is bounded by

$$CBn^2\epsilon + \frac{CB^4 n}{\epsilon^4} \log \frac{CB^4}{\epsilon^4}.$$

Taking

$$\epsilon = \left(\frac{B^3 \log n}{n} \right)^{1/5},$$

this gives the bound

$$CB^{8/5} n^{9/5} (\log n)^{1/5} \left(1 + \frac{\log B}{\log n} \right).$$

By Theorem 8.1, this completes the proof of the upper bound. The lower bound follows easily from Theorem 8.1 and the bound on $\sum c_{(ij)(ij)}$ obtained above. This finishes the proof of Theorem 8.4. $\square$

8.4 Nonlinear Large Deviations

Take a twice continuously differentiable $f : [0, 1]^N \to \mathbb{R}$, and let a, b_i, c_{ij} and $\mathcal{D}(\epsilon)$ be as in Sect. 8.1. Take any $p \in (0, 1)$ and let $Y = (Y_1, \ldots, Y_N)$ be a vector of i.i.d. Bernoulli(p) random variables. For $x \in [0, 1]$ let

$$I_p(x) := x \log \frac{x}{p} + (1 - x) \log \frac{1 - x}{1 - p}, \qquad (8.4.1)$$

and for each $x = (x_1, \ldots, x_N) \in [0, 1]^N$, let

$$I_p(x) := \sum_{i=1}^{N} I_p(x_i). \qquad (8.4.2)$$

For $t \in \mathbb{R}$, define

$$\phi_p(t) := \inf\{I_p(x) : x \in [0, 1]^N \text{ such that } f(x) \geq tN\}. \qquad (8.4.3)$$

The main result of this section, stated below, gives a sufficient condition under which the following upper tail approximation holds:

$$\mathbb{P}(f(Y) \geq tN) = \exp(-\phi_p(t) + \text{lower order terms}). \qquad (8.4.4)$$

Like the approximation (8.1.3), this is also a mean field approximation. The low complexity condition on the gradient of f comes into play, as before.

Theorem 8.5 *For f as above, $p \in (0, 1)$ and Y a vector of N i.i.d. Bernoulli(p) random variables, let ϕ_p be defined as in (8.4.3). Then, for any $\delta > 0$, $\epsilon > 0$ and $t \in \mathbb{R}$,*

$$\log \mathbb{P}(f(Y) \geq tN) \leq -\phi_p(t - \delta) + \text{complexity} + \text{smoothness},$$

where, with a, b, c_{ij}, $\mathcal{D}(\epsilon)$ defined as in Sect. 8.1, and $K := \phi_p(t)/N$,

$$\text{complexity} := \frac{1}{4}\left(N \sum_{i=1}^{N} \beta_i^2\right)^{1/2} \epsilon + 3N\epsilon + \log\left(\frac{4K(\frac{1}{N}\sum_{i=1}^{N} b_i^2)^{1/2}}{\delta \epsilon}\right)$$

$$+ \log |\mathcal{D}((\delta\epsilon)/(4K))|, \quad and$$

$$\text{smoothness} := 4\left(\sum_{i=1}^{N}(\alpha\gamma_{ii} + \beta_i^2) + \frac{1}{4}\sum_{i,j=1}^{N}(\alpha\gamma_{ij}^2 + \beta_i\beta_j\gamma_{ij} + 4\beta_i\gamma_{ij})\right)^{1/2}$$

$$+ \frac{1}{4}\left(\sum_{i=1}^{N} \beta_i^2\right)^{1/2}\left(\sum_{i=1}^{N} \gamma_{ii}^2\right)^{1/2} + 3\sum_{i=1}^{N} \gamma_{ii} + \log 2,$$

where

$$\alpha := NK + N|\log p| + N|\log(1 - p)|,$$

$$\beta_i := \frac{2Kb_i}{\delta} + |\log p| + |\log(1 - p)|, \ and$$

$$\gamma_{ij} := \frac{2Kc_{ij}}{\delta} + \frac{6Kb_ib_j}{N\delta^2}.$$

Moreover,

$$\log \mathbb{P}(f(Y) \geq tN) \geq -\phi_p(t + \delta_0) - \epsilon_0 N - \log 2,$$

where

$$\epsilon_0 := \frac{1}{\sqrt{N}}\left(4 + \left|\log \frac{p}{1 - p}\right|\right)$$

and

$$\delta_0 := \frac{2}{N}\left(\sum_{i=1}^{N}(ac_{ii} + b_i^2)\right)^{1/2}.$$

The proof of this theorem is based on Theorem 8.1. The main idea is to write the probability of $f(Y) \geq tN$ approximately as a partition function of the type considered in Theorem 8.1.

Proof (Proof of the Upper Bound in Theorem 8.5) Let $h : \mathbb{R} \to \mathbb{R}$ be a function that is twice continuously differentiable, non-decreasing, and satisfies $h(x) = -1$ if $x \leq -1$ and $h(x) = 0$ if $x \geq 0$. Let $L_1 := \|h'\|$ and $L_2 := \|h''\|$. A specific choice of h is given by

$$h(x) = 10(x + 1)^3 - 15(x + 1)^4 + 6(x + 1)^5 - 1$$

for $-1 \leq x \leq 0$, which gives $L_1 \leq 2$ and $L_2 \leq 6$. Define

$$\psi(x) := Kh((x - t)/\delta).$$

Then clearly

$$\|\psi\| \leq K, \quad \|\psi'\| \leq \frac{L_1 K}{\delta}, \quad \|\psi''\| \leq \frac{L_2 K}{\delta^2}.$$

Let

$$g(x) := N\psi(f(x)/N) + \sum_{i=1}^{N}(x_i \log p + (1 - x_i)\log(1 - p)).$$

The plan is to apply Theorem 8.1 to the function g instead of f. Note that $\psi(x) = 0$ if $x \geq t$. Thus,

$$\mathbb{P}(f(Y) \geq tN) \leq \mathbb{E}(e^{N\psi(f(Y)/N)})$$

$$= \sum_{x \in \{0,1\}^N} e^{g(x)}.$$

Note also that for any $x \in [0, 1]^N$ such that $f(x) \geq tN$,

$$g(x) - I(x) = N\psi(f(x)/N) - I_p(x) = -I_p(x) \leq -\phi_p(t).$$

Again, if $f(x) \leq (t - \delta)N$, then $(f(x)/N - t)/\delta \leq -1$, and so

$$g(x) - I(x) = -NK - I_p(x) \leq -NK = -\phi_p(t).$$

Finally, note that if $f(x) = (t - \delta')N$ for some $0 < \delta' < \delta$, then

$$g(x) - I(x) \leq -I_p(x) \leq -\phi_p(t - \delta') \leq -\phi(t - \delta).$$

Combining the three cases, we see that

$$\sup_x(g(x) - I(x)) \leq -\phi_p(t - \delta).$$

Let $C_p := |\log p| + |\log(1 - p)|$. Note that

$$\|g\| \leq NK + NC_p = \alpha,$$

and for any i,

$$\left\| \frac{\partial g}{\partial x_i} \right\| \leq \frac{2Kb_i}{\delta} + C_p = \beta_i,$$

and for any i, j,

$$\left\| \frac{\partial^2 g}{\partial x_i \partial x_j} \right\| \leq \frac{2Kc_{ij}}{\delta} + \frac{6Kb_i b_j}{N\delta^2} = \gamma_{ij}.$$

Next, fix some $\epsilon > 0$ and let $\mathscr{D}(\epsilon)$ be as in Sect. 8.1 (for the function f). Let

$$\epsilon' := \frac{\epsilon}{2\|\psi'\|}, \quad \tau := \frac{\epsilon}{2\left(\frac{1}{N}\sum_{i=1}^{N} b_i^2\right)^{1/2}}.$$

Let $l \in \mathbb{R}^N$ be the vector whose coordinates are all equal to $\log(p/(1-p))$ and define

$$\mathscr{D}'(\epsilon) := \{\theta d + l : d \in \mathscr{D}(\epsilon'),\ \theta = j\tau \text{ for some integer } 0 \le j < \|\psi'\|/\tau\}.$$

Let $g_i := \partial g/\partial x_i$. Take any $x \in [0, 1]^N$, and choose $d \in \mathscr{D}(\epsilon)$ satisfying (8.1.1). Choose an integer j between 0 and $\|\psi'\|/\tau$ such that

$$|\psi'(f(x)/N) - j\tau| \le \tau.$$

Let $d' := j\tau d + l$, so that $d' \in \mathscr{D}'(\epsilon)$. Then

$$\sum_{i=1}^{N}(g_i(x) - d_i')^2 = \sum_{i=1}^{N}(\psi'(f(x)/N)f_i(x) - j\tau d_i)^2$$

$$\le 2(\psi'(f(x)/N) - j\tau)^2\sum_{i=1}^{N}f_i(x)^2 + 2\|\psi'\|^2\sum_{i=1}^{N}(f_i(x) - d_i)^2$$

$$\le 2\tau^2\sum_{i=1}^{N}b_i^2 + 2\|\psi'\|^2 N\epsilon'^2 = N\epsilon^2.$$

This shows that $\mathscr{D}'(\epsilon)$ plays the role of $\mathscr{D}(\epsilon)$ for the function g. Note that

$$|\mathscr{D}'(\epsilon)| \le \frac{\|\psi'\|}{\tau}|\mathscr{D}(\epsilon')|.$$

This gives the upper bound on the complexity term for the function g. The proof is completed by applying Theorem 8.1. $\square$

Proof (Proof of the Lower Bound in Theorem 8.5) Fix any $z \in [0, 1]^N$ such that

$$f(z) \ge (t + \delta_0)N.$$

Let $Z = (Z_1, \ldots, Z_N)$ be a random vector with independent components, where $Z_i \sim \text{Bernoulli}(z_i)$. Let $\mathscr{A}$ be the set of all $x \in \{0, 1\}^N$ such that $f(x) \ge tN$. Let g be defined as in (8.1.17), and let $\mathscr{A}'$ be the subset of $\mathscr{A}$ where $|g(x, z) - g(x, p) - I_p(z)| \leq \epsilon_0 N$. Then

$$\mathbb{P}(f(Y) \geq tN) = \sum_{x \in \mathscr{A}} e^{g(x,p)} \tag{8.4.5}$$

$$= \sum_{x \in \mathscr{A}} e^{g(x,p) - g(x,z) + g(x,z)}$$

$$\geq \sum_{x \in \mathscr{A}'} e^{g(x,p) - g(x,z) + g(x,z)} \geq e^{-I_p(z) - \epsilon_0 N} \mathbb{P}(Z \in \mathscr{A}').$$

Note that

$$\mathbb{E}(g(Z, z) - g(Z, p)) = I_p(z),$$

and

$$\mathrm{Var}(g(Z, z) - g(Z, p))$$

$$= \sum_{i=1}^{N} \mathrm{Var}(Z_i \log(z_i/p) + (1 - Z_i) \log((1 - z_i)/(1 - p)))$$

$$= \sum_{i=1}^{N} z_i(1 - z_i) \left(\log \frac{z_i/p}{(1 - z_i)/(1 - p)} \right)^2.$$

Using the inequalities $|\sqrt{x} \log x| \leq 2/e \leq 1$ and $x(1 - x) \leq 1/4$, we see that for any $x \in [0, 1]$,

$$x(1 - x) \left(\log \frac{x/p}{(1 - x)/(1 - p)} \right)^2$$

$$\leq \left(|\sqrt{x} \log x| + |\sqrt{1 - x} \log(1 - x)| + \frac{1}{2} \left| \log \frac{p}{1 - p} \right| \right)^2$$

$$\leq \left(2 + \frac{1}{2} \left| \log \frac{p}{1 - p} \right| \right)^2.$$

Combining the last three displays, we see that

$$\mathbb{P}(|g(Z, z) - g(Z, p) - I_p(z)| > \epsilon_0 N)$$

$$\leq \frac{1}{\epsilon_0^2 N} \left(2 + \frac{1}{2} \left| \log \frac{p}{1 - p} \right| \right)^2 = \frac{1}{4}. \tag{8.4.6}$$

Let $S := f(Z) - f(z)$ and $v_i(t, x) := f_i(tx + (1 - t)z)$. Let $Z^{(i)}$ be defined following the same convention as in the proof of Theorem 8.1. Let $S_i := f(Z^{(i)}) - f(z)$, so that $|S - S_i| \le b_i$. Since

$$S = \int_0^1 \sum_{i=1}^N (Z_i - z_i) v_i(t, Z) \, dt,$$

we have

$$\mathbb{E}(S^2) = \int_0^1 \sum_{i=1}^N \mathbb{E}((Z_i - z_i) v_i(t, Z) S) \, dt. \tag{8.4.7}$$

By the independence of Z_i and the pair $(S_i, Z^{(i)})$,

$$\left| \mathbb{E}((Z_i - z_i) v_i(t, Z) S) \right|$$
$$= \left| \mathbb{E}((Z_i - z_i)(v_i(t, Z) S - v_i(t, Z^{(i)}) S_i)) \right|$$
$$\le \left\| \frac{\partial v_i}{\partial x_i} \right\| \mathbb{E}|S| + \|v_i\| \mathbb{E}|S - S_i|$$
$$\le 2atc_{ii} + b_i^2.$$

By (8.4.7), this gives

$$\mathbb{E}(S^2) \le \sum_{i=1}^N (ac_{ii} + b_i^2).$$

Therefore,

$$\mathbb{P}(f(Z) < tN) \le \frac{1}{\delta_0^2 N^2} \sum_{i=1}^N (ac_{ii} + b_i^2) = \frac{1}{4}. \tag{8.4.8}$$

Inequalities (8.4.6) and (8.4.8) give

$$\mathbb{P}(Z \in \mathscr{A}') \ge \frac{1}{2}.$$

Plugging this into (8.4.5) and taking supremum over z completes the argument. □

8.5 Quantitative Estimates for Homomorphism Densities

Let $G = G(n, p)$ be an Erdős-Rényi random graph on n vertices, with edge probability p. Let H be a fixed finite simple graph. Recall the definition of the homomorphism density $t(H, G)$ from Chap. 3. Let $\mathscr{P}_n$ and $t(H, x)$ be defined as in Sect. 8.2. Recall the function I_p defined in Eq. (8.4.1), and for $x \in \mathscr{P}_n$, let

$$I_p(x) := \sum_{1 \le i < j \le n} I_p(x_{ij}).$$

For each $u > 1$ define

$$\psi_p(u) := \inf\{I_p(x) : x \in \mathscr{P}_n \text{ such that } t(H, x) \ge u \mathbb{E}(t(H, G))\}.$$

The following theorem shows that for any $u > 1$,

$$\mathbb{P}\big(t(H, G) \ge u \mathbb{E}(t(H, G))\big) = \exp(-\psi_p(u) + \text{lower order terms}). \qquad (8.5.1)$$

provided that n is large and p is not too small. This is a quantitative version of Theorem 6.3 from Chap. 6.

Theorem 8.6 *Take any finite simple graph H and let ψ_p be defined as above. Let $X := t(H, G)$. Let k be the number of vertices and m be the number of edges of H. Suppose that $m \ge 1$ and $p \ge n^{-1/(m+3)}$. Then for any $u > 1$ and any n sufficiently large (depending only on H and u),*

$$1 - \frac{c(\log n)^{b_1}}{n^{b_2} p^{b_3}} \le \frac{\psi_p(u)}{-\log \mathbb{P}(X \ge u \mathbb{E}(X))} \le 1 + \frac{C(\log n)^{B_1}}{n^{B_2} p^{B_3}},$$

where c and C are constants that depend only on H and u, and

$$b_1 = 1, \quad b_2 = \frac{1}{2m}, \quad b_3 = 2m,$$

$$B_1 = \frac{9 + 8m}{5 + 8m}, \quad B_2 = \frac{1}{5 + 8m}, \quad B_3 = 2m - \frac{16m}{k(5 + 8m)}.$$

The remainder of this section is devoted to the proof of Theorem 8.6. The proof is a direct application of Theorem 8.5, using the estimates obtained in Sect. 8.2. The first step is to understand the properties of the rate function $\phi_p(t)$ defined in Eq. (8.4.3), put in the context of this problem. We begin with a technical lemma.

Lemma 8.7 *For any r and any $a_1, \ldots, a_r, b \in [0, 1]$,*

$$\prod_{i=1}^{r} (a_i + b(1 - a_i)) \ge (1 - b^r) \prod_{i=1}^{r} a_i + b^r.$$

Proof The proof is by induction on r. The inequality is an equality for $r = 1$. Suppose that it holds for $r - 1$. Then

$$\prod_{i=1}^{r}(a_i + b(1 - a_i)) \geq \left((1 - b^{r-1})\prod_{i=1}^{r-1} a_i + b^{r-1}\right)((1 - b)a_r + b)$$

$$= (1 - b^{r-1})(1 - b)\prod_{i=1}^{r} a_i + b^{r-1}(1 - b)a_r + (1 - b^{r-1})b\prod_{i=1}^{r-1} a_i + b^r$$

$$\geq ((1 - b^{r-1})(1 - b) + b^{r-1}(1 - b) + (1 - b^{r-1})b)\prod_{i=1}^{r} a_i + b^r$$

$$= (1 - b^r)\prod_{i=1}^{r} a_i + b^r.$$

This completes the induction.   □

Lemma 8.8 *Let $\phi_p(t)$ be defined as in (8.4.3), with $f(x) = n^2 t(H, x)$ and $N = n(n - 1)/2$. Let l be the element of $\mathscr{P}_n$ whose coordinates are all equal to 1, and let $t_0 := f(l)/N$. Then for any $0 < \delta < t < t_0$,*

$$\phi_p(t - \delta) \geq \phi_p(t) - \left(\frac{\delta}{t_0 - t}\right)^{1/m} N\log(1/p).$$

Proof Take any $x \in \mathscr{P}_n$ such that $f(x) \geq (t - \delta)N$ and x minimizes $I_p(x)$ among all x satisfying this inequality. If $f(x) \geq tN$, then we immediately have $\phi_p(t) \leq I_p(x) = \phi_p(t - \delta)$, and there is nothing more to prove. So let us assume that $f(x) < tN$. Let

$$\epsilon := \left(\frac{tN - f(x)}{f(l) - f(x)}\right)^{1/m}.$$

For each $1 \leq i < j \leq n$, let

$$y_{ij} := x_{ij} + \epsilon(1 - x_{ij}).$$

Then $y \in \mathscr{P}_n$, and by Lemma 8.7,

$$f(y) \geq (1 - \epsilon^m)f(x) + \epsilon^m f(l) = tN.$$

Thus, by the convexity of I_p,

$$\phi_p(t) \leq I_p(y) = I_p((1 - \epsilon)x + \epsilon l)$$

$$\leq (1 - \epsilon)I_p(x) + \epsilon I_p(l)$$

$$\leq I_p(x) + \epsilon N\log(1/p) = \phi_p(t - \delta) + \epsilon N\log(1/p).$$

Since $f(x) \geq (t - \delta)N$,

$$\epsilon^m \leq \frac{tN - (t - \delta)N}{f(l) - (t - \delta)N} \leq \frac{\delta}{t_0 - t}.$$

This completes the proof of the lemma.   □

Lemma 8.9 *For any p and t,*

$$\phi_p(t) \leq \frac{1}{2}(\lceil t^{1/k}n \rceil + k)^2 \log(1/p).$$

Proof Let $r := \lceil t^{1/k}n \rceil + k$. Define $x \in [0, 1]^N$ as

$$x_{ij} := \begin{cases} 1 & \text{if } 1 \leq i < j \leq r, \\ p & \text{otherwise.} \end{cases}$$

Then

$$f(x) \geq \frac{1}{n^{k-2}} \sum_{q \in [r]^k} \prod_{\{l,l'\} \in E} x_{q_l q_{l'}}$$

$$\geq \frac{r(r-1)\cdots(r-k+1)}{n^{k-2}} \geq tn^2 \geq tN,$$

and since $I_p(p) = 0$,

$$I_p(x) = \sum_{i<j} I_p(x_{ij}) \leq \frac{1}{2}r^2 \log(1/p).$$

This proves the claim.   □

The final ingredient we need before starting the proof of Theorem 8.6 is the following concentration inequality for homomorphism densities.

Lemma 8.10 *Let X be as in the statement of Theorem 8.6 and suppose that $n > k$. Then for any $u > 1$,*

$$\mathbb{P}(X \geq u\mathbb{E}(X)) \leq e^{-Cn^2 p^{2m}},$$

where C is a positive constant that depends only on u and k.

Proof Recall that X is a function of $n(n-1)/2$ i.i.d. Bernoulli(p) random variables. If the value of one of these variables changes, then from the definition of $t(H, G)$ we see that X can change by at most n^{-2}. Moreover, observe that

$$(u - 1)\mathbb{E}(X) \geq (u - 1)n(n - 1) \cdots (n - k + 1)n^{-k}p^m,$$

since any map that takes the vertices of H to distinct vertices of G has a probability p^m of being a homomorphism. The claim now follows easily by McDiarmid's inequality (Theorem 4.3).   □

We are now ready to prove Theorem 8.6.

Proof (Proof of the Upper Bound in Theorem 8.6) The task now is to pull together all the information obtained above, for use in Theorem 8.5. As before, we work with $f(x) = n^2 t(H,x)$. Take $t = \kappa p^m$ for some fixed $\kappa > 0$. Let δ and ϵ be two positive real numbers, both less than t, to be chosen later. Note that $\delta < t < \kappa p^{2m/k}$ since $t = \kappa p^m$ and $k > 2$. Assume that δ and ϵ are bigger than $n^{-1/2}$. Note that p is already assumed to be bigger than $n^{-1/2}$ in the statement of the theorem.

Recall that the indexing set for quantities like b_i and c_{ij}, instead of being $\{1,\ldots,N\}$, is now $\{(i,j) : 1 \le i < j \le n\}$. For simplicity, we will write (ij) instead of (i,j). Throughout, C will denote any constant that depends only on the graph H, the constant κ, and nothing else. From Theorem 8.2, we have the estimates

$$a \le n^2, \quad b_{(ij)} \le C,$$

and

$$c_{(ij)(i'j')} \le \begin{cases} Cn^{-1} & \text{if } |\{i,j,i',j'\}| = 2 \text{ or } 3, \\ Cn^{-2} & \text{if } |\{i,j,i',j'\}| = 4. \end{cases}$$

Let $\theta := \delta^{-1} p^{2m/k}$. By Lemma 8.9,

$$K \le C p^{2m/k} \log n.$$

Using the above bounds, we get

$$\alpha \le Cn^2 \log n, \quad \beta_{(ij)} \le C\theta \log n,$$

and

$$\gamma_{(ij)(i'j')} \le \begin{cases} Cn^{-1}\theta \log n & \text{if } |\{i,j,i',j'\}| = 2 \text{ or } 3, \\ Cn^{-2}\delta^{-1}\theta \log n & \text{if } |\{i,j,i',j'\}| = 4. \end{cases}$$

Therefore, we have the estimates

$$\sum_{(ij)} \beta_{(ij)}^2 \le Cn^2\theta^2(\log n)^2, \quad \sum_{(ij)} b_{(ij)}^2 \le Cn^2,$$

and by Theorem 8.3,

$$\log |\mathscr{D}((\delta\epsilon)/(4K))| \leq \frac{Cn\theta^4}{\epsilon^4} \log \frac{CK}{\delta\epsilon}$$

$$\leq \frac{Cn\theta^4(\log n)^5}{\epsilon^4}.$$

Combining the last three estimates, we see that the complexity term in Theorem 8.5 is bounded above by

$$Cn^2\epsilon\theta \log n + \frac{Cn\theta^4(\log n)^5}{\epsilon^4}.$$

Taking $\epsilon = n^{-1/5}\theta^{3/5}(\log n)^{4/5}$, the above bound simplifies to

$$Cn^{9/5}\theta^{8/5}(\log n)^{9/5}.$$

Next, note that by the bounds obtained above and the inequality $\delta > n^{-1/2}$,

$$\sum_{(ij)} \alpha\gamma_{(ij)(ij)} \leq Cn^3\theta(\log n)^2,$$

$$\sum_{(ij),(i'j')} \alpha\gamma^2_{(ij)(i'j')} \leq Cn^3\theta^2(\log n)^3,$$

$$\sum_{(ij),(i'j')} \beta_{(ij)}(\beta_{(i'j')} + 4)\gamma_{(ij)(i'j')} \leq Cn^2\delta^{-1}\theta^3(\log n)^3,$$

$$\left(\sum_{(ij)} \beta^2_{(ij)}\right)^{1/2} \left(\sum_{(ij)} \gamma^2_{(ij)(ij)}\right)^{1/2} \leq Cn\theta^2(\log n)^2,$$

$$\sum_{(ij)} \gamma_{(ij)(ij)} \leq Cn\theta \log n.$$

The above estimates show that the smoothness term in Theorem 8.5 is bounded above by a constant times

$$n^{3/2}\theta(\log n)^{3/2} + n\delta^{-1/2}\theta^{3/2}(\log n)^{3/2} + n\theta^2(\log n)^2.$$

Putting $\eta := p^{2m/k}$, we see that this is bounded by a constant times

$$n^{3/2}\delta^{-1}\eta(\log n)^{3/2} + n\delta^{-2}\eta^{3/2}(\log n)^2.$$

Since $\delta > n^{-1/2}$, we can further simplify this upper bound to

$$n^{3/2}\delta^{-1}\eta(\log n)^2.$$

Combining the bounds on the complexity term and the smoothness term, we get that

$$\log \mathbb{P}(f(Y) \geq tN) \leq -\phi_p(t - \delta) + Cn^{9/5}\delta^{-8/5}\eta^{8/5}(\log n)^{9/5}$$
$$+ Cn^{3/2}\delta^{-1}\eta(\log n)^2.$$

By Lemma 8.8,

$$-\phi_p(t - \delta) \leq -\phi_p(t) + C\delta^{1/m}n^2 \log n.$$

Taking

$$\delta = n^{-m/(5+8m)}\eta^{8m/(5+8m)}(\log n)^{4m/(5+8m)}$$

gives

$$\log \mathbb{P}(f(Y) \geq tN) \hspace{4cm} (8.5.2)$$
$$\leq -\phi_p(t) + Cn^{(9+16m)/(5+8m)}\eta^{8/(5+8m)}(\log n)^{(9+8m)/(5+8m)}$$
$$+ Cn^{(15+26m)/(10+16m)}\eta^{5/(5+8m)}(\log n)^{(10+12m)/(5+8m)}.$$

Now note that since $p > n^{-1/2}$,

$$\frac{n^{(9+16m)/(5+8m)}\eta^{8/(5+8m)}}{n^{(15+26m)/(10+16m)}\eta^{5/(5+8m)}} = n^{(3+6m)/(10+16m)}p^{6m/(5k+8mk)}$$
$$\geq n^{(3+6m)/(10+16m)}n^{-3m/(5+8m)}$$
$$= n^{3/(10+16m)}.$$

This shows that the first term on the right-hand side in (8.5.2) dominates the second when n is sufficiently large. Therefore, when n is large enough,

$$\log \mathbb{P}(f(Y) \geq tN)$$
$$\leq -\phi_p(t) + Cn^{(9+16m)/(5+8m)}p^{16m/(5k+8mk)}(\log n)^{(9+8m)/(5+8m)}.$$

Written differently, this is

$$\frac{\phi_p(t)}{-\log \mathbb{P}(f(Y) \geq tN)}$$

$$\leq 1 + \frac{Cn^{(9+16m)/(5+8m)}p^{16m/(5k+8mk)}(\log n)^{(9+8m)/(5+8m)}}{-\log \mathbb{P}(f(Y) \geq tN)}.$$

By Lemma 8.10,

$$-\log \mathbb{P}(f(Y) \geq tN) \geq Cn^2 p^{2m}, \qquad (8.5.3)$$

Therefore,

$$\frac{\phi_p(t)}{-\log \mathbb{P}(f(Y) \geq tN)}$$

$$\leq 1 + Cn^{-1/(5+8m)}p^{-2m+16m/(5k+8mk)}(\log n)^{(9+8m)/(5+8m)}.$$

A minor verification using the assumption $p \geq n^{-1/4}$ shows that the ϵ and δ chosen above are both bigger than $n^{-1/2}$, as required. To complete the proof of the upper bound, notice that $\mathbb{E}(X)$ is asymptotic to p^m since $p \geq n^{-1/(m+3)}$.   □

Proof (Proof of the Lower Bound in Theorem 8.6) First, observe that by Lemma 8.8, Theorem 8.2, and the lower bound in Theorem 8.5,

$$\log \mathbb{P}(f(Y) \geq tN) \geq -\phi_p(t) - Cn^{-1/2m}n^2 \log n.$$

Therefore, again applying (8.5.3), we get

$$\frac{\phi_p(t)}{-\log \mathbb{P}(f(Y) \geq tN)} \geq 1 - Cn^{-1/2m}p^{-2m}\log n.$$

This completes the proof of the lower bound.   □

8.6   Explicit Rate Function for Triangles

Theorem 8.6 converts the large deviation question for subgraph densities into a question of solving a variational problem analogous to the variational problem for dense graphs that we saw in Theorem 5.2. Unlike the dense case, however, the sparse regime allows us to explicitly solve the variational problem in many examples. The simplest example, that of triangle density, is solved in this section. We will prove the following result.

Theorem 8.7 *Let $T_{n,p}$ be the number of triangles in an Erdős–Rényi graph $G(n,p)$. Then for any fixed $\delta > 0$, as $n \to \infty$ and $p \to 0$ simultaneously such that $p \gg n^{-1/158}(\log n)^{33/158}$,*

$$\mathbb{P}(T_{n,p} \geq (1+\delta)\mathbb{E}(T_{n,p}))$$

$$= \exp\left(-(1-o(1))\min\left\{\frac{\delta^{2/3}}{2}, \frac{\delta}{3}\right\} n^2 p^2 \log(1/p)\right).$$

Let H be a triangle, and for simplicity, let $T(f) := t(H,f)$. Let $\mathscr{P}_n$ and $t(H,x)$ be as in Sect. 8.2. For $x \in \mathscr{P}_n$, let $T(x) := t(H,x)$. For $\delta > 0$, let

$$\psi_p(n,\delta) := \inf\{I_p(x) : x \in \mathscr{P}_n, \ T(x) \geq (1+\delta)p^3\}.$$

By Theorem 8.6, it suffices to show that under the given conditions,

$$\frac{\psi_p(n,\delta)}{n^2 p^2 \log(1/p)} \to \min\left\{\frac{\delta^{2/3}}{2}, \frac{\delta}{3}\right\}. \tag{8.6.1}$$

The remainder of this section is devoted to the proof of (8.6.1). The proof requires a number of technical estimates, which are worked out below. Throughout, we will use the notation $a \ll b$ to mean that the two quantities a and b are varying in such a way that $a/b \to 0$. Similarly, $a \sim b$ will mean that $a/b \to 1$. The symbol $o(a)$ will denote any quantity that, when divided by a, tends to zero as the relevant parameters vary in some prescribed manner. Let I_p be defined as in (8.4.1).

Lemma 8.11 *If $p \to 0$ and $0 \leq x \ll p$, then $I_p(p+x) \sim x^2/(2p)$. On the other hand, if $p \to 0$, $x \to 0$ and $p \ll x$, then $I_p(p+x) \sim x\log(x/p)$.*

Proof Note that $I_p(p) = I_p'(p) = 0$, $I_p''(p) = 1/(p(1-p))$ and

$$I_p'''(x) = \frac{1}{(1-x)^2} - \frac{1}{x^2}. \tag{8.6.2}$$

The first assertion of the lemma follows easily by Taylor expansion, using the above identities. For the second assertion, note that if $p \to 0$, $x \to 0$ and $p \ll x$, then

$$I_p(p+x) = (p+x)\log\frac{p+x}{p} + (1-p-x)\log\frac{1-p-x}{1-p}$$

$$= x\log\frac{x}{p} + O(x).$$

This completes the proof of the lemma. $\square$

Lemma 8.12 *There exists $p_0 > 0$ such that for all $0 < p \leq p_0$ and $0 \leq x \leq b \leq$*
$1 - p - 1/\log(1/p)$,

$$I_p(p + x) \geq (x/b)^2 I_p(p + b).$$

Proof Let $q := 1 - p - 1/\log(1/p)$. Let $f(t) := I_p(p + \sqrt{t})$. Note that

$$f''(t) = \frac{d}{dt}\left(\frac{1}{2\sqrt{t}}I_p'(p + \sqrt{t})\right)$$

$$= \frac{1}{4t}I_p''(p + \sqrt{t}) - \frac{1}{4t^{3/2}}I_p'(p + \sqrt{t}).$$

Let $g(x) := 4x^3 f''(x^2)$. Then by the above formula,

$$g(x) = xI_p''(p + x) - I_p'(p + x).$$

By (8.6.2), this shows that

$$g'(x) = xI_p'''(p + x) \begin{cases} \leq 0 & \text{if } p + x \leq 1/2, \\ \geq 0 & \text{if } p + x \geq 1/2. \end{cases} \tag{8.6.3}$$

Now note that $g(0) = 0$, and as $p \to 0$,

$$g(q) = \frac{q\log(1/p)}{(1 - 1/\log(1/p))} - \log\frac{(1 - p)(\log(1/p) - 1)}{p}$$

$$= -\log\log(1/p) + O(1).$$

Thus, there exists p_0 small enough such that if $p \leq p_0$, then $g(q) < 0$. By (8.6.3), this implies that $g(x) \leq 0$ for all $x \in [0, q]$. In particular, f is concave in the interval $[0, q^2]$. Thus, for any $0 \leq x \leq b \leq q$,

$$I_p(p + x) = f(x^2) \geq (x^2/b^2)f(b^2) = (x/b)^2 I_p(p + b),$$

which completes the proof.   □

Corollary 8.1 *There is some $p_0 > 0$ such that when $0 < p \leq p_0$ and $0 \leq x \leq 1-p$,*

$$I_p(p + x) \geq x^2 I_p(1 - 1/\log(1/p)).$$

Proof Let p_0 be as in Lemma 8.12 and let $b = 1-p-1/\log(1/p)$. When $0 \leq x \leq b$, the inequality follows by Lemma 8.12 since $b < 1$. When $b \leq x \leq 1 - p$, then we trivially deduce $I_p(p + x) \geq I_p(p + b) \geq x^2 I_p(p + b)$.   □

Now let

$$\phi_p(\delta) := \frac{1}{2}\inf\{I_p(f) : f \in \mathcal{W}, T(f) \geq (1+\delta)p^3\}.$$

Note that unlike $\psi_p(n,\delta)$, $\phi_p(\delta)$ does not depend on n. The following lemma compares $\psi_p(n,\delta)$ and $\phi_p(\delta)$.

Lemma 8.13 *For any p, n and* δ,

$$\phi_p(\delta) \leq \frac{1}{n^2}\psi_p(n,\delta) + \frac{1}{2n}\log\frac{1}{1-p}.$$

Proof For any $x \in \mathscr{P}_n$, construct a graphon f in the natural way, by defining $f(s,t) = x_{ij}$ if either (s,t) or (t,s) belongs to the square $[(i-1)/n, i/n) \times [(j-1)/n, j/n)$ for some $1 \leq i \leq j \leq n$, where x_{ii} is taken to be zero. It is easy to verify that $T(f) = T(x)$, and

$$n^2 I_p(f) = 2I_p(x) + nI_p(0).$$

Dividing by $2n^2$ on both sides and taking infimum over x completes the proof. □
We are now ready to prove Theorem 8.7.

Proof (Proof of Theorem 8.7) Throughout this proof, C will denote any universal constant and $o(1)$ will denote any quantity that depends only on p and tends to zero as $p \to 0$.
   Take any f such that $T(f) \geq (1+\delta)p^3$. First, we wish to show that

$$\frac{1}{2}I_p(f) \geq (1-o(1))\min\left\{\frac{\delta^{2/3}}{2}, \frac{\delta}{3}\right\}p^2\log(1/p). \tag{8.6.4}$$

By Lemma 8.13, this will show that

$$\liminf\frac{\psi_p(n,\delta)}{n^2p^2\log(1/p)} \geq \min\left\{\frac{\delta^{2/3}}{2}, \frac{\delta}{3}\right\},$$

proving one direction of (8.6.1). Assume that

$$\frac{1}{2}I_p(f) \leq \min\left\{\frac{\delta^{2/3}}{2}, \frac{\delta}{3}\right\}p^2\log(1/p). \tag{8.6.5}$$

We can make this assumption because if there exists no such f (among all f satisfying $T(f) \geq (1+\delta)p^3$), then the proof of (8.6.4) is automatically complete.
   Since I_p is decreasing in $[0,p]$ and increasing in $[p,1]$, we may assume without loss of generality that $f \geq p$ everywhere. (Otherwise, it suffices to prove (8.6.4)

for $f' := \max\{f, p\}$, because $T(f') \geq T(f) \geq (1 + \delta)p^3$ and $I_p(f') \leq I_p(f)$.) Let $g := f - p$. Then

$$T(f) = T(p + g) = p^3 + T(g) + 3pR(g) + 3p^2 S(g), \qquad (8.6.6)$$

where

$$R(g) := \int_{[0,1]^3} g(x, y)g(y, z)\, dx\, dy\, dz$$

and

$$S(g) := \int_{[0,1]^2} g(x, y)\, dx\, dy.$$

Now, if $S(g) \geq p^{3/2}\log(1/p)$, then by Lemma 8.11,

$$I_p(f) = I_p(p + g) \geq I_p(p + S(g))$$
$$\geq I_p(p + p^{3/2}\log(1/p)) \gg p^2 \log(1/p),$$

which proves (8.6.4). So let us assume that

$$S(g) < p^{3/2}\log(1/p). \qquad (8.6.7)$$

Under the above assumption, (8.6.6) gives

$$T(g) + 3pR(g) \geq (\delta - o(1))p^3. \qquad (8.6.8)$$

For each $x \in [0, 1]$, let

$$d(x) := \int_0^1 g(x, y)\, dy,$$

and for each $b \in [0, 1]$, let

$$B_b := \{x \in [0, 1] : d(x) \geq b\}.$$

Note that by Jensen's inequality,

$$I_p(f) = I_p(p + g) \geq \int_0^1 I_p(p + d(x))\, dx.$$

Thus,

$$\mathrm{Leb}(B_b) \leq \frac{\int_0^1 I_p(p + d(x))\, dx}{I_p(p + b)} \leq \frac{I_p(f)}{I_p(p + b)}.$$

Therefore by the assumption (8.6.5), we get

$$\mathrm{Leb}(B_b) \leq \frac{Cp^2 \log(1/p)}{I_p(p+b)}.$$

If b varies with p such that $p \ll b < 1-p$, then by Lemma 8.11, the above inequality gives

$$\mathrm{Leb}(B_b) \leq \frac{Cp^2 \log(1/p)}{b \log(b/p)} \leq \frac{Cp^2}{b}. \tag{8.6.9}$$

Thus,

$$\int_{B_b \times B_b} g(x,y)^2 \, dx \, dy \leq \mathrm{Leb}(B_b)^2 \leq \frac{Cp^4}{b^2}. \tag{8.6.10}$$

Let D be the set of all $(x,y,z) \in [0,1]^3$ such that at least one of x, y and z belongs to B_b. Choose b depending on p such that $\sqrt{p \log(1/p)} \ll b \ll 1$. Then by (8.6.7) and (8.6.9),

$$\int_D g(x,y)g(y,z)g(z,x) \, dx \, dy \, dz \leq 3\mathrm{Leb}(B_b)S(g) \ll p^3. \tag{8.6.11}$$

Let $B_b^c := [0,1] \backslash B_b$, and let

$$\delta_1 := \frac{1}{p^2} \int_{B_b \times B_b^c} g(x,y)^2 \, dx \, dy$$

and

$$\delta_2 := \frac{1}{p^2} \int_{B_b^c \times B_b^c} g(x,y)^2 \, dx \, dy.$$

By the inequality (8.6.11) and the generalized Hölder inequality from Chap. 2 (Theorem 2.1),

$$T(g) = \int_{B_b^c \times B_b^c \times B_b^c} g(x,y)g(y,z)g(z,x) \, dx \, dy \, dz + o(p^3)$$

$$\leq (\delta_2^{3/2} + o(1))p^3. \tag{8.6.12}$$

Next, by Lemma 8.12 and Jensen's inequality,

$$I_p(f) = I_p(p + g) \geq \int_0^1 I_p(p + d(x))\, dx$$

$$\geq \int_{B_b^c} I_p(p + d(x))\, dx$$

$$\geq \frac{I_p(p + b)}{b^2} \int_{B_b^c} d(x)^2\, dx.$$

Proceeding as in the proof of (8.6.9), this gives

$$\int_{B_b^c} d(x)^2\, dx \leq \frac{b^2 I_p(f)}{I_p(p + b)} \leq Cp^2 b = o(p^2).$$

Therefore,

$$R(g) = \int_0^1 d(x)^2\, dx = \int_{B_b} d(x)^2\, dx + o(p^2)$$

$$\leq \int_{B_b} \int_0^1 g(x, y)^2\, dy\, dx + o(p^2).$$

Combining this with Eq. (8.6.10), we get

$$R(g) \leq (\delta_1 + o(1))p^2. \tag{8.6.13}$$

By Eqs. (8.6.8), (8.6.12) and (8.6.13), we get

$$3\delta_1 + \delta_2^{3/2} \geq \delta - o(1).$$

Finally, by Corollary 8.1,

$$I_p(f) = I_p(p + g) \geq (1 - o(1))(2\delta_1 + \delta_2)p^2 \log(1/p),$$

where the $o(1)$ term appears because

$$I_p(1 - 1/\log(1/p)) = (1 - o(1))\log(1/p).$$

Combining all of the above, we get

$$\liminf_{p \to 0} \frac{\phi_p(\delta)}{p^2 \log(1/p)} \geq \inf\left\{ \delta_1 + \frac{1}{2}\delta_2 : \delta_1 \geq 0,\ \delta_2 \geq 0, \right.$$

$$\left. 3\delta_1 + \delta_2^{3/2} \geq \delta \right\}. \tag{8.6.14}$$

Now take any $a \geq 0$, and for $0 \leq x \leq a/3$, let

$$f(x) := x + \frac{1}{2}(a - 3x)^{2/3}.$$

An easy verification shows that depending on the value of a, either f is increasing throughout the interval $[0, a/3]$, or increases up to a maximum value and then decreases. In either case, the minimum value of f is achieved at one of the two endpoints of the interbal $[0, a/3]$. This shows that under the constraint $3\delta_1 + \delta_2^{3/2} = a$, the minimum value of $\delta_1 + \delta_2/2$ is attained at either $\delta_1 = 0$ or $\delta_2 = 0$. Since this is true for any a, the minimization in (8.6.14) is attained when either $\delta_1 = 0$ or $\delta_2 = 0$. From this, (8.6.4) follows easily.

Let us now prove the other direction of (8.6.1), that is,

$$\limsup \frac{\psi_p(n, \delta)}{n^2 p^2 \log(1/p)} \leq \min \left\{ \frac{\delta^{2/3}}{2}, \frac{\delta}{3} \right\}. \tag{8.6.15}$$

First let a be an integer, to be chosen later, and let $x \in \mathscr{P}_n$ be the element defined as $x_{ij} = 1$ if $1 \leq i < j \leq a$, and $x_{ij} = p$ otherwise. Then it is easy to see that

$$T(x) \geq \frac{1}{n^3} (a(a-1)(a-2) + (n(n-1)(n-2) - a(a-1)(a-2))p^3)$$

and

$$I_p(x) = \frac{a(a-1)}{2} \log(1/p).$$

Thus, we can choose a such that $a = (\delta^{1/3} + o(1))np$, $T(x) \geq (1 + \delta)p^3$, and

$$I_p(x) = \left(\frac{\delta^{2/3}}{2} + o(1) \right) n^2 p^2 \log(1/p).$$

Next, let x be defined as $x_{ij} = 1$ if $1 \leq i \leq a$ and $i < j$, and $x_{ij} = p$ otherwise. Then

$$T(x) \geq \frac{1}{n^3} (3a(n-a)(n-a-1)p + (n-a)(n-a-1)(n-a-2)p^3)$$

and

$$I_p(x) = a\left(n - \frac{a+1}{2} \right) \log(1/p).$$

Thus, we can choose a such that $a = (\delta/3 + o(1))p^2 n$, $T(x) \geq (1 + \delta)p^3$ and

$$I_p(x) = \left(\frac{\delta}{3} + o(1)\right)n^2 p^2 \log(1/p).$$

Combining the two bounds proves (8.6.15). □

Bibliographical Notes

The theory developed in the earlier chapters applies only to dense random graphs. However, most graphs that arise in applications are sparse. The main roadblock to developing the analogous theory for sparse graphs is the unavailability of a suitable sparse version of graph limit theory. Although a number of attempts have been made towards extending graph limit theory to the sparse setting such as in Bollobás and Riordan [2] and Borgs et al. [3, 4], an effective version of a sparse 'counting lemma' that connects the cut metric with subgraph counts, is still missing.

There are numerous powerful upper bounds for tail probabilities of nonlinear functions of independent Bernoulli random variables, such as the bounded difference inequality popularized by McDiarmid [18], and the inequalities of Talagrand [19], Latała [16], Kim and Vu [14, 15], Vu [20] and Janson et al. [13]. However, these upper bounds generally do not yield the right constants in the exponent, which is the goal of large deviations theory. Even establishing the right powers of n and p in the rate functions for subgraph densities has been quite challenging—see, for example, Chatterjee [7] and DeMarco and Kahn [10, 11].

Motivated by the problem of understanding the large deviations of subgraph densities in sparse random graphs, the nonlinear large deviation theory presented in this chapter was developed in Chatterjee and Dembo [8]. The technique used for proving Lemmas 8.1 and 8.2 was originally developed in Chatterjee [5, 6] and Chatterjee and Dey [9] in the context of developing rigorous formulations and proofs of mean field equations. The move from mean field equations to the mean field approximation of Theorem 8.1 required the low complexity gradient condition, which was the main new idea in Chatterjee and Dembo [8]. The proof of Theorem 8.6 presented here is slightly shorter than the one in [8], eliminating the need for using a result of Janson et al. [13], which would have required a large number of additional pages if I attempted to present the proof here. This comes at the cost of a slightly worse error bound.

Theorem 8.7 was proved in Lubetzky and Zhao [17]. The proof presented here is almost a verbatim copy of the proof in Lubetzky and Zhao [17]. The error bounds in Chatterjee and Dembo [8] allow p to go to zero slower than $n^{-1/42}$ in Theorem 8.7. The slightly worse error bound of Theorem 8.6 is the reason why p needs to decay slower than $n^{-1/158}$ here. However, these are all very far from the conjectured threshold: It is believed that Theorem 8.7 holds whenever $p \to 0$ slower than $n^{-1/2}$. Incidentally, Lubetzky and Zhao [17] show that with the use of a weak version of

Szemerédi's regularity lemma, it is possible to probe a sparse large deviation result as long as p tends to zero slower than a negative power of $\log n$.

Lubetzky and Zhao [17] also proved analogous results for the density of cliques. These results were extended to general homomorphism densities by Bhattacharya et al. [1], who obtained the following beautiful explicit formula for the rate function for general homomorphism densities. Take any finite simple graph H with maximum degree Δ. Let H^* be the induced subgraph of H on all vertices whose degree in H is Δ. Recall that an independent set in a graph is a set of vertices such that no two are connected by an edge. Also, recall that a graph is called regular if all its vertices have the same degree, and irregular otherwise. Define a polynomial

$$P_{H^*}(x) := \sum_k i_{H^*}(k)x^k,$$

where $i_{H^*}(k)$ is the number of k-element independent sets in H^*.

Theorem 8.8 (Bhattacharya et al. [1])   *Let H be a connected finite simple graph on k vertices with maximum degree $\Delta \geq 2$. Then for any $\delta > 0$, there is a unique positive number $\theta = \theta(H, \delta)$ that solves $P_{H^*}(\theta) = 1 + \delta$, where P_{H^*} is the polynomial defined above. Let $H_{n,p}$ be the number of homomorphisms of H into an Erdős–Rényi $G(n, p)$ random graph. Then there is a constant $\alpha_H > 0$ depending only on H, such that if $n \to \infty$ and $p \to 0$ slower than $n^{-\alpha_H}$, then for any $\delta > 0$,*

$$\mathbb{P}(H_{n,p} \geq (1 + \delta)\mathbb{E}(H_{n,p})) = \exp\left(-(1 + o(1))c(\delta)n^2 p^\Delta \log \frac{1}{p} \right),$$

where

$$c(\delta) = \begin{cases} \min\{\theta, \frac{1}{2}\delta^{2/k}\} & \text{if H is regular,} \\ \theta & \text{if H is irregular.} \end{cases}$$

The formula given in Theorem 8.8 is more than just a formula. It gives a hint at the conditional structure of the graph, and at the nature of phase transitions as δ varies. Unlike the dense case, it is hard to give a precise meaning to claims about the conditional structure in the sparse setting due to the lack of an adequate sparse graph limit theory.

The paper [1] also gives a number of examples where the coefficient $c(\delta)$ in Theorem 8.8 can be explicitly computed. For instance, if $H = C_4$, the cycle of length four, then

$$c(\delta) = \begin{cases} \frac{1}{2}\sqrt{\delta} & \text{if $\delta < 16$,} \\ -1 + \sqrt{1 + \frac{1}{2}\delta} & \text{if $\delta \geq 16$.} \end{cases}$$

Theorem 8.7 is also a special case of Theorem 8.8.

A very interesting recent development in nonlinear large deviations is the preprint of Eldan [12]. Eldan's proof technique is quite different than the one presented here. Eldan's approach is also based on the low complexity gradient condition, although the complexity of the gradient is measured differently, as follows. Recall that the Gaussian width of a measurable set $K \subseteq \mathbb{R}^N$ is defined as

$$\mathrm{GW}(K) := \mathbb{E}(\sup_{x \in K} x \cdot Z),$$

where Z denotes a standard Gaussian random vector and $x \cdot Z$ denotes the inner product of x and Z. Eldan measure the complexity of the gradient of a function $f : \{0, 1\}^N \to \mathbb{R}$ as

$$\mathscr{D}(f) := \mathrm{GW}(\{\Delta f(x) : x \in \{0, 1\}^N\} \cup \{0\}),$$

where $\Delta f(x) = (\Delta_1 f(x), \ldots, \Delta_N f(x))$ is the discrete gradient of f, with $\Delta_i f(x)$ defined as in (8.1.4). Eldan measures the smoothness of f using the Lipschitz constant $\mathrm{Lip}(f)$ of f with respect to the Hamming distance on $\{0, 1\}^N$. In terms of the discrete derivatives, $\mathrm{Lip}(f) = \max_{i,x} |\Delta_i f(x)|$. One of the main results of Eldan's paper is the following improved version of Theorem 8.5.

Theorem 8.9 (Eldan [12]) *Let all notation be as in Theorem 8.5. Let $\mathrm{Lip}(f)$ denote the Lipschitz constant of f with respect to the Hamming distance on $\{0, 1\}^N$. Then for any $t, \delta \in \mathbb{R}$ such that $0 < \delta < \phi_p(t - \delta)/n$,*

$$\log \mathbb{P}(f(Y) \geq tN) \leq -\phi_p(t - \delta)(1 - 6\delta^{-1}ABN^{-1/3}(\log N)^{1/6}),$$

where

$$A = \left(\mathrm{Lip}(f) + \frac{1}{48\sqrt{n}} \mathrm{Lip}(f)^2 + |\log(p(1-p))| \right)^{2/3},$$

$$B = \left(2\mathscr{D}(f) + \frac{1}{48} \mathrm{Lip}(f)^2 \right)^{1/3}.$$

Moreover, if $\mathrm{Lip}(f)^2 \leq n\delta^2$, then

$$\log \mathbb{P}(f(Y) \geq tN) \geq -\phi_p(t)\left(1 + \frac{1}{2n\delta^2} \mathrm{Lip}(f)^2 \right) - \log 10.$$

The quantities A and B correspond to the smoothness and complexity terms in Theorem 8.5. Theorem 8.9 has several advantages over Theorem 8.5. First, it has a cleaner statement. Second, it does not use the second derivatives of f at all. Third, it gives improved error bounds. For example, using Theorem 8.9, Eldan [12] demonstrated that Theorem 8.7 holds whenever $p \to 0$ slower than $n^{-1/18}$. This, unfortunately, is still far from the conjectured optimal threshold of $n^{-1/2}$.

Lastly, another interesting recent development is the paper of Yan [21], which extends the nonlinear large deviation theory to functions of a more general class of random variables than Bernoulli. In particular, Yan's theorems apply to continuous random variables.

References

1. Bhattacharya, B. B., Ganguly, S., Lubetzky, E., & Zhao, Y. (2015). Upper tails and independence polynomials in random graphs. *arXiv preprint arXiv:1507.04074*.
2. Bollobás, B., & Riordan, O. (2009). Metrics for sparse graphs. In *Surveys in combinatorics 2009*. London Mathematical Society Lecture Note Series (vol. 365, pp. 211–287). Cambridge: Cambridge Univeristy Press.
3. Borgs, C., Chayes, J. T., Cohn, H., & Zhao, Y. (2014). An L^p theory of sparse graph convergence I: Limits, sparse random graph models, and power law distributions. *arXiv preprint arXiv:1401.2906*.
4. Borgs, C., Chayes, J. T., Cohn, H., & Zhao, Y. (2014). An L^p theory of sparse graph convergence II: LD convergence, quotients, and right convergence. *arXiv preprint arXiv:1408.0744*.
5. Chatterjee, S. (2005). Concentration inequalities with exchangeable pairs. Ph.D. thesis, Stanford University.
6. Chatterjee, S. (2007). Stein's method for concentration inequalities. *Probability Theory and Related Fields, 138*(1–2), 305–321.
7. Chatterjee, S. (2012). The missing log in large deviations for triangle counts. *Random Structures & Algorithms, 40*(4), 437–451.
8. Chatterjee, S., & Dembo, A. (2016). Nonlinear large deviations. *Advances in Mathematics, 299*, 396–450.
9. Chatterjee, S., & Dey, P. S. (2010). Applications of Stein's method for concentration inequalities. *Annals of Probability, 38*, 2443–2485.
10. DeMarco, B., & Kahn, J. (2012). Upper tails for triangles. *Random Structures & Algorithms, 40*(4), 452–459.
11. DeMarco, B., & Kahn, J. (2012). Tight upper tail bounds for cliques. *Random Structures & Algorithms, 41*(4), 469–487.
12. Eldan, R. (2016). Gaussian-width gradient complexity, reverse log-Sobolev inequalities and nonlinear large deviations. *arXiv preprint arXiv:1612.04346*.
13. Janson, S., Oleszkiewicz, K., & Ruciński, A. (2004). Upper tails for subgraph counts in random graphs. *Israel Journal of Mathematics, 142*, 61–92.
14. Kim, J. H., & Vu, V. H. (2000). Concentration of multivariate polynomials and its applications. *Combinatorica, 20*(3), 417–434.
15. Kim, J. H., & Vu, V. H. (2004). Divide and conquer martingales and the number of triangles in a random graph. *Random Structures & Algorithms, 24*(2), 166–174.
16. Latała, R. (1997). Estimation of moments of sums of independent real random variables. *Annals of Probability, 25*(3), 1502–1513.
17. Lubetzky E., & Zhao, Y. (2017). On the variational problem for upper tails of triangle counts in sparse random graphs. *Random Structures & Algorithms, 50*(3), 420–436.
18. McDiarmid, C. (1989). On the method of bounded differences. In J. Siemons (Ed.), *Surveys in combinatorics*. London Mathematical Society Lecture Note Series (vol. 141, pp. 148–188). Cambridge: Cambridge Univeristy Press.
19. Talagrand, M. (1995). Concentration of measure and isoperimetric inequalities in product spaces. *Publications Mathématiques de l'Institut des Hautes Études Scientifiques, 81*, 73–205.
20. Vu, V. H. (2002). Concentration of non-Lipschitz functions and applications. Probabilistic methods in combinatorial optimization. *Random Structures & Algorithms, 20*(3), 262–316.
21. Yan, J. (2017). Nonlinear large deviations: Beyond the hypercube. *arXiv preprint arXiv:1703.08887*.

Index

© Springer International Publishing AG 2017
S. Chatterjee, *Large Deviations for Random Graphs*, Lecture Notes
in Mathematics 2197, DOI 10.1007/978-3-319-65816-2

LECTURE NOTES IN MATHEMATICS  Springer

Editors in Chief: J.-M. Morel, B. Teissier;

Editorial Policy

1. Lecture Notes aim to report new developments in all areas of mathematics and their applications – quickly, informally and at a high level. Mathematical texts analysing new developments in modelling and numerical simulation are welcome.

    Manuscripts should be reasonably self-contained and rounded off. Thus they may, and often will, present not only results of the author but also related work by other people. They may be based on specialised lecture courses. Furthermore, the manuscripts should provide sufficient motivation, examples and applications. This clearly distinguishes Lecture Notes from journal articles or technical reports which normally are very concise. Articles intended for a journal but too long to be accepted by most journals, usually do not have this "lecture notes" character. For similar reasons it is unusual for doctoral theses to be accepted for the Lecture Notes series, though habilitation theses may be appropriate.

2. Besides monographs, multi-author manuscripts resulting from SUMMER SCHOOLS or similar INTENSIVE COURSES are welcome, provided their objective was held to present an active mathematical topic to an audience at the beginning or intermediate graduate level (a list of participants should be provided).

    The resulting manuscript should not be just a collection of course notes, but should require advance planning and coordination among the main lecturers. The subject matter should dictate the structure of the book. This structure should be motivated and explained in a scientific introduction, and the notation, references, index and formulation of results should be, if possible, unified by the editors. Each contribution should have an abstract and an introduction referring to the other contributions. In other words, more preparatory work must go into a multi-authored volume than simply assembling a disparate collection of papers, communicated at the event.

3. Manuscripts should be submitted either online at www.editorialmanager.com/lnm to Springer's mathematics editorial in Heidelberg, or electronically to one of the series editors. Authors should be aware that incomplete or insufficiently close-to-final manuscripts almost always result in longer refereeing times and nevertheless unclear referees' recommendations, making further refereeing of a final draft necessary. The strict minimum amount of material that will be considered should include a detailed outline describing the planned contents of each chapter, a bibliography and several sample chapters. Parallel submission of a manuscript to another publisher while under consideration for LNM is not acceptable and can lead to rejection.

4. In general, **monographs** will be sent out to at least 2 external referees for evaluation.

    A final decision to publish can be made only on the basis of the complete manuscript, however a refereeing process leading to a preliminary decision can be based on a pre-final or incomplete manuscript.

    Volume Editors of **multi-author works** are expected to arrange for the refereeing, to the usual scientific standards, of the individual contributions. If the resulting reports can be forwarded to the LNM Editorial Board, this is very helpful. If no reports are forwarded or if other questions remain unclear in respect of homogeneity etc, the series editors may wish to consult external referees for an overall evaluation of the volume.

5. Manuscripts should in general be submitted in English. Final manuscripts should contain at least 100 pages of mathematical text and should always include

   - a table of contents;
   - an informative introduction, with adequate motivation and perhaps some historical remarks: it should be accessible to a reader not intimately familiar with the topic treated;
   - a subject index: as a rule this is genuinely helpful for the reader.
   - For evaluation purposes, manuscripts should be submitted as pdf files.

6. Careful preparation of the manuscripts will help keep production time short besides ensuring satisfactory appearance of the finished book in print and online. After acceptance of the manuscript authors will be asked to prepare the final LaTeX source files (see LaTeX templates online: https://www.springer.com/gb/authors-editors/book-authors-editors/manuscriptpreparation/5636) plus the corresponding pdf- or zipped ps-file. The LaTeX source files are essential for producing the full-text online version of the book, see http://link.springer.com/bookseries/304 for the existing online volumes of LNM). The technical production of a Lecture Notes volume takes approximately 12 weeks. Additional instructions, if necessary, are available on request from lnm@springer.com.

7. Authors receive a total of 30 free copies of their volume and free access to their book on SpringerLink, but no royalties. They are entitled to a discount of 33.3 % on the price of Springer books purchased for their personal use, if ordering directly from Springer.

8. Commitment to publish is made by a *Publishing Agreement*; contributing authors of multiauthor books are requested to sign a *Consent to Publish form*. Springer-Verlag registers the copyright for each volume. Authors are free to reuse material contained in their LNM volumes in later publications: a brief written (or e-mail) request for formal permission is sufficient.

Addresses:
Professor Jean-Michel Morel, CMLA, École Normale Supérieure de Cachan, France
E-mail: moreljeanmichel@gmail.com

Professor Bernard Teissier, Equipe Géométrie et Dynamique,
Institut de Mathématiques de Jussieu – Paris Rive Gauche, Paris, France
E-mail: bernard.teissier@imj-prg.fr

Springer: Ute McCrory, Mathematics, Heidelberg, Germany,
E-mail: lnm@springer.com

Printed in the United States
By Bookmasters

Printed in the United States
By Bookmasters